Dual Pandemics

Dual Pandemics: Creating Racially-Just Responses to a Changing Environment through Research, Practice and Education commits to promoting and disseminating knowledge that calls for the dismantling of systemic racism and creating racially-just responses to the dual pandemics.

COVID-19 and anti-racist uprisings as a result of the murders of Mr. George Floyd and many other African Americans and other people of color due to police violence have unprecedented impact on our society. While these two pandemics appear to be different in nature, both pandemics attest to the fact that systemic racism continues to be a grand challenge and that COVID-19 differentially affects communities and people of color as well as socially disadvantaged groups. This book offers intellectually sound examination, conceptualization, and rigor in providing viable, socially just, and responsive paths forward. The volume includes chapters that focus on anti-racist pedagogy in social work education, conceptual discussion contributing to refining a shared understanding of constructs relevant to anti-racist social work, and micro, mezzo, and macro social work practice that aims to prevent or eliminate the negative impact of racism as well as promote racial justice, equity, and inclusion among individuals, families, groups, organizations, or communities.

This book will be of great value to students and scholars of social work, public policy, race, and ethnic studies. The chapters in this book were originally published as a special issue of the *Journal of Ethnic & Cultural Diversity in Social Work*.

Mo Yee Lee is Professor and PhD Program Director, College of Social Work, The Ohio State University. She is editor of the *Journal of Ethnic & Cultural Diversity in Social Work* since 2007 and president of The Group for the Advancement of Doctoral Education in Social Work from 2019 to 2021.

Monit Cheung is Professor of Social Work, Mary R. Lewis Endowed Professor in Children & Youth, Principal Investigator of Child Welfare Education Project, and Director of Child & Family Center for Innovative Research at the Graduate College of Social Work, University of Houston, Houston, Texas.

Michael A. Robinson is Professor at the University of Georgia School of Social Work. He is the author of over 40 articles and book chapters. Two of his articles have received national awards. He has focused his research on the unjust killings of African American males by police and continues to publish in this area.

Michele Rountree is Associate Professor and Academy of Distinguished Teacher with the Steve Hicks School of Social Work at the University of Texas at Austin. She founded Black Mamas ATX, providing culturally aligned, client-centered, and mental health services that support Black women through pregnancy and postpartum period.

Michael S. Spencer (Kānaka Maoli) is Presidential Term Professor at the University of Washington School of Social Work and Director of Native Hawaiian, Pacific Islander, and Oceania Affairs at the Indigenous Wellness Research Institute. He is Grand Challenges for Social Work Executive Committee Member and co-lead for the Close the Health Gap and Eliminate Racism Grand Challenges.

Martell L. Teasley is Interim Senior Vice President of Academic Affairs and Provost. Appointed dean of the College of Social Work in 2017, Martell is in his second term as President of the National Association of Deans and Directors of Social Work. As the lead investigator on the Social Work profession's Grand Challenge to Eliminate Racism, his major areas of research interest are African American adolescent development, school social work practice, and diversity in social work education.

Dual Pandemics

Creating Racially-Just Responses to a Changing
Environment through Research, Practice
and Education

Edited by
**Mo Yee Lee, Monit Cheung, Michael A. Robinson,
Michele Rountree, Michael S. Spencer
and Martell L. Teasley**

Routledge
Taylor & Francis Group

LONDON AND NEW YORK

First published 2024
by Routledge
4 Park Square, Milton Park, Abingdon, Oxon OX14 4RN

and by Routledge
605 Third Avenue, New York, NY 10158

Routledge is an imprint of the Taylor & Francis Group, an informa business

Introduction, Chapters 2–15 © 2024 Taylor & Francis
Chapter 1 © 2022 Andrea Murray-Lichtman, Adriana Aldana, Elena Izaksonas, Tauchiana Williams, Mitra Naseh, Anne C. Deepak and Michele Rountree. Originally published as Open Access.

British Library Cataloguing in Publication Data
A catalogue record for this book is available from the British Library

ISBN13: 978-1-032-37208-2 (hbk)
ISBN13: 978-1-032-37210-5 (pbk)
ISBN13: 978-1-003-33585-6 (ebk)

DOI: 10.4324/9781003335856

Typeset in Minion Pro
by Newgen Publishing UK

Publisher's Note
The publisher accepts responsibility for any inconsistencies that may have arisen during the conversion of this book from journal articles to book chapters, namely the inclusion of journal terminology.

Disclaimer
Every effort has been made to contact copyright holders for their permission to reprint material in this book. The publishers would be grateful to hear from any copyright holder who is not here acknowledged and will undertake to rectify any errors or omissions in future editions of this book.

Contents

Citation Information

The chapters in this book were originally published in the *Journal of Ethnic & Cultural Diversity in Social Work*, volume 31, issue 3–5 (2022). When citing this material, please use the original page numbering for each article, as follows:

Introduction

Dual pandemics: creating racially-just responses to a changing environment through research, practice and education
Mo Yee Lee, Monit Cheung, Michael A. Robinson, Michele Rountree, Michael Spencer, and Martell L. Teasley
Journal of Ethnic & Cultural Diversity in Social Work, volume 31, issue 3–5 (2022), pp. 135–138

Chapter 1

Dual pandemics awaken urgent call to advance anti-racism education in social work: pedagogical illustrations
Andrea Murray-Lichtman, Adriana Aldana, Elena Izaksonas, Tauchiana Williams, Mitra Naseh, Anne C. Deepak, and Michele A. Rountree
Journal of Ethnic & Cultural Diversity in Social Work, volume 31, issue 3–5 (2022), pp. 139–150

Chapter 2

Advancing critical race pedagogical approaches in social work education
Patrina Duhaney, Liza Lorenzetti, Kaltrina Kusari, and Emily Han
Journal of Ethnic & Cultural Diversity in Social Work, volume 31, issue 3–5 (2022), pp. 151–161

Chapter 3

Coloniality of power, critical realism and critical consciousness: the three "C" framework
Lisa Werkmeister Rozas
Journal of Ethnic & Cultural Diversity in Social Work, volume 31, issue 3–5 (2022), pp. 162–172

Chapter 4

Anti-racism and equity-mindedness in social work field education: a systematic review
Candice C. Beasley, Melissa I. Singh, and Katherine Drechsler
Journal of Ethnic & Cultural Diversity in Social Work, volume 31, issue 3–5 (2022), pp. 173–185

Chapter 13

Mask mandates, race, and protests of summer 2020

Rahbel Rahman, Sameena Azhar, Laura J. Wernick, Jordan E. DeVylder, Tina Maschi, Margaret Cohen, and Simone Hopwood

Journal of Ethnic & Cultural Diversity in Social Work, volume 31, issue 3–5 (2022), pp. 280–291

Chapter 14

Model Minority Mutiny: addressing anti-Asian racism during the COVID-19 pandemic in social work

Dale Dagar Maglalang, Smitha Rao, Bongki Woo, and Kaipeng Wang

Journal of Ethnic & Cultural Diversity in Social Work, volume 31, issue 3–5 (2022), pp. 292–301

Chapter 15

Conceptualizing anti-Asian racism in Canada during the COVID-19 pandemic: A call for action to social workers

Kedi Zhao, Carolyn O'Connor, Trish Lenz, and Lin Fang

Journal of Ethnic & Cultural Diversity in Social Work, volume 31, issue 3–5 (2022), pp. 302–312

For any permission-related enquiries please visit:
www.tandfonline.com/page/help/permissions

Notes on Contributors

Adriana Aldana, California State University, Dominguez Hills, California, USA.

Sameena Azhar, Fordham University Graduate School of Social Service, New York, New York, USA.

Candice C. Beasley, School of Social Work, Tulane University, New Orleans, Louisiana, USA.

Stephanie Boddie, Diana R. Garland School of Social Work, the School of Education, and the George W. Truett Theological Seminary, Baylor University, USA; College of Human Sciences, University of South Africa, Pretoria, South Africa; Institute for Gender Studies, Pretoria, South Africa.

Kristen Brock-Petroshius, Luskin School of Public Affairs, Department of Social Welfare, University of California, Los Angeles, California, USA.

Mimi Chapman, School of Social Work, University of North Carolina, Chapel Hill, North Carolina, USA.

Monit Cheung, Graduate College of Social Work, University of Houston, Houston, Texas, USA.

Margaret Cohen, Graduate School of Social Service, Fordham University, New York, New York, USA.

Fernanda Cross, University of Michigan School of Social Work, Ann Arbor, Michigan, USA.

Anne C. Deepak, School of Social Work, Monmouth University, West Long Branch, New Jersey, USA.

Jordan E. DeVylder, Graduate School of Social Service, Fordham University, New York, New York, USA.

Katherine Drechsler, School of Social Work, University of Wisconsin-Whitewater, Whitewater, Wisconsin, USA.

Patrina Duhaney, Faculty of Social Work, University of Calgary, Calgary, Alberta, Canada.

Lin Fang, Factor-Inwentash Faculty of Social Work, University of Toronto, Toronto, Ontario, Canada.

Sarah Godoy, School of Social Work, University of North Carolina, Chapel Hill, North Carolina, USA.

Trenette Clark Goings, School of Social Work, University of North Carolina, Chapel Hill, North Carolina, USA.

Odessa Gonzalez Benson, University of Michigan School of Social Work, Ann Arbor, Michigan, USA.

Rachel W. Goode, School of Social Work, University of North Carolina, Chapel Hill, North Carolina, USA

David Halpern, University Center for Social and Urban Research, University of Pittsburgh, Pittsburgh, Pennsylvania, USA.

Emily Han, Faculty of Social Work, University of Calgary, Calgary, Alberta, Canada.

Simone Hopwood, Graduate School of Social Service, Fordham University, New York, New York, USA.

Kimberly D. Hudson, Fordham University Graduate School of Social Service, New York, New York, USA.

Elena Izaksonas, Metropolitan State University, Saint Pauls, Minnesota, USA.

Kirk James, Silver School of Social Work, New York University, New York, New York, USA.

Sarah Johansson, Graduate School of Social Service, Fordham University, New York, New York, USA.

Winnie W. Kung, Graduate School of Social Service, Fordham University, New York, New York, USA.

Kaltrina Kusari, Faculty of Social Work, University of Calgary, Calgary, Alberta, Canada.

Eric Kyere, Indiana University School of Social Work, Indiana University – Purdue University Indianapolis, Indianapolis, Indiana, USA.

Jessica, Euna Lee, Indiana University School of Social Work, Indiana University – Purdue University Indianapolis, Indianapolis, Indiana, USA.

Mo Yee Lee, College of Social Work, The Ohio State University, Columbus, Ohio, USA.

Trish Lenz, Factor-Inwentash Faculty of Social Work, University of Toronto, Toronto, Ontario, Canada.

Liza Lorenzetti, Faculty of Social Work, University of Calgary, Calgary, Alberta, Canada.

Dale Dagar Maglalang, Center for Alcohol and Addiction Studies, Brown University School of Public Health, Providence, Rhode Island, USA.

Tina Maschi, Graduate School of Social Service, Fordham University, New York, New York, USA.

Elizabeth B. Matthews, Fordham University Graduate School of Social Service, New York, New York, USA.

Dominique Mikell, Luskin School of Public Affairs, Department of Social Welfare, University of California, Los Angeles, California, USA.

Christopher Sanjurjo Montalvo, University of Michigan School of Social Work, Ann Arbor, Michigan, USA.

Andrea Murray-Lichtman, Chapel Hill School of Social Work, University of North Carolina, Chapel Hill, North Carolina, USA.

Mitra Naseh, School of Social Work, Portland State University, Portland, Oregon, USA.

Carolyn O'Connor, Factor-Inwentash Faculty of Social Work, University of Toronto, Toronto, Ontario, Canada.

Ebony Nicole Perez, Undergraduate Social Work, Saint Leo University, Saint Leo, Florida, USA.

Rahbel Rahman, Fordham University Graduate School of Social Service, New York, New York, USA.

Smitha Rao, College of Social Work, the Ohio State University, Columbus, Ohio, USA.

Michael A. Robinson, School of Social Work, University of Georgia, Athens, Georgia, USA.

Abigail M. Ross, Fordham University Graduate School of Social Service, New York, New York, USA.

Michele Rountree, Steve Hicks School of Social Work, University of Texas at Austin, USA.

Kevan Schultz, University Center for Social and Urban Research, University of Pittsburgh, Pittsburgh, Pennsylvania, USA.

Caroline N. Sharkey, School of Social Work, University of Georgia, Athens, Georgia, USA.

Melissa I. Singh, Suzanne Dworak-Peck School of Social Work, University of Southern California, Los Angeles, California, USA.

Michael S. Spencer, School of Social Work, University of Washington, Seattle, WA, USA.

Christopher A. Strickland, School of Social Work, University of Georgia, Athens, Georgia, USA.

Martell L. Teasley, College of Social Work, The University of Utah, Salt Lake City, Utah, USA.

Kaipeng Wang, Graduate School of Social Work, University of Denver Denver, Colorado, USA.

Durrell Malik Washington Sr., Crown Family School of Social Work, Policy, and Practice, University of Chicago, Chicago, Illinois, USA.

Lisa Werkmeister Rozas, School of Social Work, University of Connecticut, Storrs, Connecticut, USA.

Laura J. Wernick, Graduate School of Social Service, Fordham University, New York, New York, USA.

Tauchiana Williams, Chapel Hill School of Social Work, University of North Carolina, Chapel Hill, North Carolina, USA.

Bongki Woo, College of Social Work, University of South Carolina, Columbia, South Carolina, USA.

Kedi Zhao, Factor-Inwentash Faculty of Social Work, University of Toronto, Toronto, Ontario, Canada.

Introduction—Dual pandemics: creating racially-just responses to a changing environment through research, practice and education

Mo Yee Lee, Monit Cheung, Michael A. Robinson, Michele Rountree, Michael S. Spencer and Martell L. Teasley

2020 is plagued by two pandemics: COVID-19 & anti-racist uprisings as a result of the murders of Mr. George Floyd and many other African Americans and other people of color due to police violence. While these two pandemics appear to be different in nature, both pandemics attest to the fact that systemic racism continues to be a grand challenge in our society. It is appalling to see how COVID-19 differentially affects communities and people of color as well as socially disadvantaged groups. Systemic racism and police brutality are related to the unacceptable violation of human rights of diverse groups in the US and globally.

The present moment provides distinct challenges and opportunities for the promise of the social work profession. Social work has long distinguished itself as a profession ethically bound to stand for social justice. Generations of emerging social workers matriculate in accredited social work programs to uphold this value in service to and with those populations regulated to the margins who have disproportionately suffered from the dual pandemics of COVID-19 and police violence. With years of experience as social work researchers and educators, the editorial board views the special issue as representing the profession's values of confronting systemic racism. The articles are timely from social work scholars and educators in a time where the threat, at least in one state of the proposed legislation would fire tenured faculty for teaching tenets of critical race theory. Knowledge becomes politicized in the context of the dual pandemics. There has been intense tension across state legislation and schools on whether discrimination and bias should be addressed in education. Many states across the United States have proposed bills to control CRT in schools: https://www.edweek.org/policy-politics/map-where-critical-race-theory-is-under-attack/2021/06.

Refusing to be deterred, the authors of special issue articles offer intellectually sound examination, conceptualization, and rigor in providing viable, socially just, responsive paths forward. This Special Issue includes articles that focus on anti-racist pedagogy in social work education, conceptual discussion contributing to refining a shared understanding of constructs relevant to anti-racist social work, and micro, mezzo, and macro social work practice that aims to prevent or eliminate the negative impact of racism as well as promote racial justice, equity, and inclusion among individuals, families, groups, organizations, or communities. Mental, physical health, and safety should not be considered utopian values. nor dignity and grace of all individuals.

Part I – anti-racist pedagogy in social work education

The special issue includes five articles that refine the constructs and examine the practice of anti-racist pedagogy in social work education. "Dual pandemics awaken urgent call to advance anti-racism education in social work: Pedagogical illustrations" describes the efforts by members of the

CSWE Task Force for Anti-racism. The discussion describes a path that promotes racial justice, dismantles systemic racism, and eliminates white supremacy within social work education. This article interrogates social work's complicity in white supremacy, provides examples of social work anti-racism pedagogy, and calls for centering BIPOC voices to move social work toward its anti-racism future.

"Advancing critical race pedagogical approaches in social work education" addresses the lack of research-informed transformative learning models in social work that has resulted in the persistent centering of Western ways of knowing in Canada and globally. The authors propose critical race pedagogy as an essential framework to promote and enrich social work learning environments where students can engage in courageous conversations about race, racism, power, and oppression.

Reflecting on the dual pandemics, COVID-19 and racism in the United States and the fact that typical pedagogical practices tend to center on Whiteness when teaching about issues of race, racism, and racial justice, "Coloniality of Power, Critical Realism and Critical Consciousness" further expands the discussion and proposes a framework that de-centers the White frame using three C's: the coloniality of power, critical realism, and critical consciousness. This framework intentionally promotes a pedagogical foundation that highlights accountability, agency, and self-emancipation.

To provide a picture of what has been done in social work pedagogy, "Anti-Racism and Equity-Mindedness in Social Work Field Education: A Systematic Review" synthesizes the literature over the last ten years to examine field education programs and critically examines anti-racist and equity-mindedness in field education, which is the signature pedagogy in social work education.

"Lift Ev'ry Voice: Social Work Educators' Experiences Teaching Race and Racism" explores the experiences of undergraduate social work educators who taught the required diversity courses. This qualitative study highlights the primary challenges of teaching about race and racism, including (1) faculty racial identity and lack of credibility and (2) the emotional toll of teaching about race. Study findings have important implications concerning educators' preparation to engage in anti-racist and equitable pedagogy in undergraduate social work education programs.

Part II – conceptualizing anti-racist social work practice and research

Five articles focus on conceptual discussions that contribute to refining or expanding a shared understanding of constructs relevant to anti-racist social work. The article, "Dual pandemics or a syndemic? Racism, COVID-19, and opportunities for antiracist social work," critically examines the conceptualization of racism as a pandemic and the limitations of medicalizing racism. The authors introduce the term syndemic and discuss how the language of syndemics might accurately characterize the synergism and interconnectedness of racism and COVID-19 and the potential of a syndemic theory in offering insights and opportunities for social work research, practice, and policy from a racial justice lens.

"Visualizing Structural Competency: Moving Beyond Cultural Competence/ Humility Toward Eliminating Racism" is another conceptual discussion that examines racism from a macro, structural perspective. Building on structural violence as a theoretical framework and the need for social work to explicitly build structural competency to effectively respond to structural racism and respond to the Grand Challenge to Eliminate Racism, this article presents a "structuragram" as a heuristic to assess, analyze, and intervene at the structural level factors that influence the individual and community realities in addition to including a case example and recommendations for structural competency-based practice.

In the article, "From Social Justice to Abolition: Living Up to Social Work's Grand Challenge of Eliminating Racism," the authors argue for the replacement of the predominant social justice paradigm with a framework for anti-racist social work praxis informed by abolitionist principles that emphasize the building of power in Black, Indigenous, or Brown and poor communities. The discussion shares concrete anti-racist praxis tools and defines praxis principles that include engaging with critical theories, advancing macro-approaches, targeting racism at the source, and developing interventions to eliminate and address the effects of racism.

"Power Knowledge in Social Work: Educating Social Workers to Practice Racial Justice" analyzes and interrogates knowledge-production practices in contemporary social work research and practice through the lens of Michel Foucault's concept of power-knowledge. As a regime of power, social work produces forms of knowledge that stratify human subjects along with the social fabric. To reconcile a contemporary social work professional logic saturated in white supremacy with a longstanding ethical mandate for social justice, this investigation concludes with practice and pedagogical recommendations informed by an anti-racist theoretical framework.

Integrating Beauchamp and Childress's four ethical principles as an overarching framework with NASW's code of ethics and the Grand Challenges for Social Work on eliminating racism, "Ethical Mental Health Practice in Diverse Cultures and Races" examines their intersection with cultural diversity and anti-racism and its implications for mental health services. This article highlights the importance of inclusivity beyond individual clients, revisiting the relevance of evidence-based practices in collaboration with clients as embedded in their culture. In addition, it urges practitioners to reduce biases, cultural ignorance, and microaggressions to avoid misdiagnosis and invalidation of client situations and ensure equitable access to culturally sensitive services for the marginalized communities.

Part III – impact of dual pandemics on special groups and populations

Dual pandemics assert differential and negative experience to vulnerable and racially diverse populations. Five articles highlight research and scholarship that centers race as a key variable and examine the impact of systemic racism and COVID-19 on different communities and groups. Two articles focus on the experience of two different groups of essential workers: Black women and migrant workers. Using a phenomenological research design, "Necessary, Yet Mistreated: The Lived Experiences of Black Women Essential Workers in Dual Pandemics of Racism and COVID-19" explores the experiences of 22 Black women essential workers navigating these dual pandemics. Findings of the study highlight their experiences related to their desire to and fear of protest; navigating extreme emotions; mixed levels of understanding from colleagues; and a rise in blatantly racist confrontations in the workplace. Another group of essential workers negatively affected by the dual pandemics is the migrant and immigrant workers. "Demanding Migrant/Immigrant Labor in the Coronavirus Crisis: Critical Perspectives for Social Work Practice" adopts a labor-focused framework as a critical perspective that complements the rights-based, participatory frameworks with immigrants to account for systemic racism, global and national inequality, and discrimination embedded in immigration and social policies and forms of practice.

Mask Mandates, Race, and Protests of Summer 2020 examined predictors to mask mandate support and racial justice protest participation across Asian (n = 103), Black (n = 102), white (n = 102) New York City residents, using binary logistic regressions. Participants with positive feelings about the racial justice movement were more likely to participate in the protests. White and Asian respondents were more likely to support the mask mandates than Black respondents. Asian respondents were less likely to participate in public protests than white respondents. Our findings offer a model for social workers to understand how race, political participation, and COVID-19 intersect to create racially just responses to health and justice matters.

Two articles examine anti-Asian racism during the dual pandemics in the United States and Canada, respectively. "Model Minority Mutiny: Addressing Anti-Asian Racism During the COVID-19 Pandemic in Social Work" critically examines the condition of Asians and Asian Americans in the United States with the goal of supporting inter-solidarity movements in social work to uplift the lived experiences of Asian Americans. The discussion introduces four recommendations pertaining to conceptualizing and positioning the Asian American identity, acknowledging the heterogeneity of the Asian American population, integrating Asian American history in the social work curriculum, and using research strategies to address anti-Asian racism.

"Conceptualizing Anti-Asian racism in Canada during the COVID-19 pandemic: A call for action to social workers" integrates Canadian postcolonialism, Canadian multiculturalism, and a framework of intergroup prejudice to conceptualize the covert anti-Asian racism that is entrenched in Canadian society. This conceptualization highlights social workers' leading roles in combating anti-Asian racism through reforming and integrating client interventions, cultural policy, and social context, and offers directions that can guide future social work research and practice in improving social justice during this crisis.

This Special Issue also includes a review of the book *Caste: The Origins of Our Discontents* by Isabel Wilkerson. This book provides a description of the caste system that has been quietly infiltrating the United States since the emergence of our nation as well as a refreshing framework to ignite new understanding, conversations, and action toward reimagining the United States as a country that is equitable for all.

Concluding thoughts

As the COVID-19 pandemic makes fewer headlines and we see the potential for normalcy in our daily lives, systemic racism continues to rage unabated. The dual pandemics thus also demonstrate the responsiveness of society when all lives are at stake compared to the uneven and performative responses to racism when BIPOC lives are at stake. Will social work respond similarly, or will the profession rise to the challenge of systemic change for racial justice?

The articles in this special issue suggest that social work is prepared to respond and that the events of 2020 will continue to stand as a moment for opportunity and growth as a profession. Many white Americans were suddenly "waken" in 2020 and had new insight into why San Francisco 49ers quarterback Colin Kaepernick took a knee and risked his lucrative football career years back. They also began to realize that "serve and protect" was not applied equally to all people. Following these events, the 13th Grand Challenge in Social Work to Eliminate Racism was established. It is uncertain if this would have occurred had the COVID-19 pandemic and the murder of George Floyd and others not happened. As we seek answers to how we move forward, social work must continue to critically challenge existing systems, including our own profession and the institutions we interact with and, in some cases, rely upon for resources. The articles presented in this special issue illustrate both the damage to BIPOC communities because of the dual pandemics, as well as ways we can make the dream of racial justice a reality.

As society stays vigilant during the current pandemic against new viral strains that put the most vulnerable at risk, social work must remain vigilant against existing forms of systemic racism as well as new mutations that will likely arise as systems work to maintain their power. This special issue serves as a call for the profession to go beyond cultural competency and develop structural competency through anti-racist research, pedagogy, policies, and practice. Consistent with systems of foundational change, we can expect our work to continue to evolve and respond to the changing dynamics of racism and oppression that will have tremendous impacts on social work research, practice, and education.

Disclosure statement

No potential conflict of interest was reported by the author(s).

Part I

Anti-racist pedagogy in social work education

Dual pandemics awaken urgent call to advance anti-racism education in social work: pedagogical illustrations

Andrea Murray-Lichtman, Adriana Aldana, Elena Izaksonas, Tauchiana Williams, Mitra Naseh, Anne C. Deepak and Michele A. Rountree

ABSTRACT

In 2020 racial justice uprisings and COVID-19 and the push for institutional responses created pressure within social work to answer decades of calls for anti-racism action. CSWE responded and formed the Task Force for Anti-racism. As members of the Task Force, we call on CSWE to continue this anti-racism work. We describe a path forward to promote racial justice and dismantle systemic racism and white supremacy within social work education. We interrogate social work's complicity in white supremacy, provide examples of social work anti-racism pedagogy, and call for centering BIPOC voices to move social work toward its anti-racism future.

Socio-political factors propel the hardships of the COVID-19 pandemic, resulting in outsized impacts on Black, Indigenous, People of Color (BIPOC) communities who are at greater risk for infection, hospitalization, and the likelihood of death. BIPOC are doubly or, in some cases, triple times over-represented. The disproportional ratios by race and ethnicity are magnified when comparing risk for infection, hospitalization, and the likelihood of death for Indigenous 1.7, 3.5, 2.4, Blacks 1.1, 2.8, 2.0, Latinos 1.9, 2.8, 2.3, and Asians .07, 1.0, 1.0 respectively, in comparison to Whites (Centers for Disease Control (CDC), 2021b). The convergence of the positionality of race, experiences of racism in the United States, and the disproportionate rates of COVID-19 present significant risk for dismal physical results, psychological trauma, and economic instability for BIPOC communities. Hence, the dual pandemics of racial justice uprisings and COVID-19 in 2020 and the ensuing push for institutional responses created pressure within social work to answer decades of calls for critical self-examination and change (Maylea, 2021).

The disproportionate pandemic related illnesses, hospitalizations, and deaths of BIPOC parents and caregivers underscore a sense of urgency to advance anti-racism in social work education and practice with BIPOC children, youth, and families. The dual pandemics exacerbate the following existing social inequities: parents working in essential low wage employment positions (Economic Policy Institute (EPI), 2020; Hawkins, 2020; Waltenburg et al., 2020), placing them at risk for exposure with limitations on preventative strategies such as social distancing or mandatory wearing of masks, uninsured or underinsured health care coverage (EPI, 2020), institutional racism (Khazanchi et al. (2020), social determinants of health (Karaye & Horney, 2020; Okoh et al., 2020; Rodriguez-Lonear et al., 2020), underlying and predisposed health conditions (CDC, 2021a), the digital divide in telemedicine (hotspots; Moore et al., 2020), interface with other institutional settings such as immigration or correctional facilities (Tobolowsky et al., 2020; Wallace et al., 2020) and overcrowded geographic areas (Millett et al., 2020). Furthermore, BIPOC children grieving the loss of a parent or caregiver are at a heightened risk for long-term physical and mental health concerns (Hillis et al., 2021).

The National Association of Black Social Workers in 1968 (Brice & McLane-Davison, 2020) and the 1973 report from the CSWE Black Task Force (Harty, 2021) voiced the first demand to address racism within social work. Spurred by the dual pandemics, the mounting pressure of BIPOC and other marginalized voices and their allies renewed demands for social work education and practice to move from individualistic non-racist stances to actions that eradicate white supremacy and systemic racism within social work. The Council of Social Work Education (CSWE, 2021) responded to this pressure and formed the 2020 CSWE Task Force for Anti-racism. As a subgroup of the 2020 CSWE Task Force for Anti-racism, we call on the profession and CSWE to examine the historical and current perpetuations of racism within social work (Beck, 2019; Corley & Young, 2018; Dominelli, 1997), eradicate manifestations of white supremacy within social work education and practice, and honor the contributions of BIPOC social workers and activists who have led this work from the inception of the profession.

The pandemic disproportionately impacted BIPOC children, youth, and families and magnified inequities in education, health, housing, and immigration. This paper describes a path forward to promote racial justice and dismantle systemic racism and white supremacy within social work education. First, we introduce operational definitions for white supremacy, anti-racism, and anti-racism pedagogy to create a shared understanding and counter controversial, misaligned interpretations. We utilize the lens of Critical Race Theory (CRT) as an anti-racism theoretical framework to inform a shared understanding of anti-racism social work pedagogy and to move anti-racism in social work beyond reflexivity to action. Second, we critique social work practice and policy with children, youth, and families to examine the impact of white supremacy in our profession's history. Third, we employ pedagogical examples to provide social work educators tools for explicit anti-racism pedagogy. We also highlight the leadership of BIPOC social work scholars, educators, and practitioners to counter the impact of racism through practices grounded in the strengths of BIPOC families, children, and communities. We conclude with pedagogical recommendations for advancing anti-racism in social work education.

Anti-racism in social work pedagogy

The amplified calls for justice provoked by COVID-19 and racial injustice (CSWE, 2021) require that we interrogate the racism underlying many theories driving social work practice and our approaches to instruction as a profession. Race as a social construct has been historically used to create and justify laws, policies, and practices that dehumanize children, youth, families, and communities and harm them through state-sponsored violence and separation from each other (Aldana & Vazquez, 2020). Historically, the social work profession and social workers have been complicit in these acts (Ioakimidis & Trimikliniotis, 2020; Maylea, 2021), and the profession has ignored and silenced the voices of diverse groups of social workers, including BIPOC–in essence, racialized populations who have contributed theory, practice, and movements to humanize, heal and liberate oppressed people. Dominelli (1997) challenges the profession to recognize "the epistemological base and political philosophy of social work education endorses the status quo, of which racism is an integral feature" (p. 43).

The Anti-racism Task Force adopted a definition of anti-racism as an active practice, process, and policy contesting racial inequality and dismantling white supremacy that involves the unlearning and challenging of the myth of white supremacy and understanding how it drives other forms of oppression. It rejects the inferiority of BIPOC to center multiple ways of being and knowing. In addition, "anti-racism is the active process of identifying and eliminating racism by changing systems, organizational structures, policies and practices and attitudes, so that power is redistributed and shared equitably" (NAC International Perspectives: Women and Global Solidarity [NAC], n.d.) through collaborative efforts inside and outside the profession to build networks of solidarity (IFSW, as cited by CSWE, 2021). White supremacy is defined as a political, economic, and cultural system in which whites overwhelmingly control power and material resources, conscious and unconscious ideas of white superiority and entitlement are widespread; relations of white dominance and nonwhite subordination are reenacted daily across a broad array of institutions and social settings (Ansley, 1989, as cited by CSWE, 2021).

Anti-racism in social work education requires a pedagogical commitment to social transformation inside and outside the classroom, starting with intensive critical thinking and self-reflection by students and instructors regarding their social positions and working toward building anti-racism networks of solidarity inside and outside the profession to eliminate racism. Anti-racism action requires that social work education examine social work's historical and contemporary narrative. Mainstream social work education erases the contributions of BIPOC social work scholars, educators, practitioners, and organizations, and we must counter this impact through policy, practice, and interventions that are grounded in the strengths of BIPOC communities, families, and children (Brice & McLane-Davison, 2020). Though seemingly benign, this epistemological erasure is an assault on racialized communities.

Integrating critical race theory into anti-racism pedagogy

One of the goals of anti-racism pedagogy is to proactively work toward "transformation by challenging the individual as well as the structural system that perpetuates racism" (Blakeney, 2005, p. 20). The emphasis of anti-racism pedagogy includes:

> challenging assumptions and fostering students' critical analytical skills; developing students' awareness of their social positions; decenter authority in the classroom and having students take responsibility for their learning process; empowering students and applying theory to practice, and creating a sense of community through collaborative learning. (Kishimoto, 2018, p. 546)

Kishimoto's ideas support the anti-racism pedagogy principles described by Hassouneh (2006); students' assumptions are challenged as they are educated in ways that make racialized power relations explicit. In addition, critical analytical skills are fostered as students deconstruct and analyze interlocking systems of oppression.

Aligned with these anti-racism ideas and principles, Critical Race Theory (CRT) offers social work education a pathway to anti-racism pedagogy. Moreover, CRT's origin in American legal scholarship provides a conceptual understanding of how race and racism intersect and inform the law and social policies; this is particularly relevant to anti-racism in social work practice and policy (Razack & Jeffery, 2002). Employing the work of thought leaders such as Frederick Douglas, W.E.B. Du Bois, bell hooks, and Toni Morrison, in the 1980s critical legal scholars Derrick Bell, Alan Freeman, Richard Delgado, Kimberle' Crenshaw, Mari Matsuda, and others generated a theory that interrogates the centrality of race and the pervasiveness of racism in systemic social inequities (Bell, 1995; Crenshaw et al., 1995). While tenets of CRT have been coopted and parts of the theory used without centering race and racism, the historical struggle of BIPOC and anti-racism action formed the foundations of CRT and have lessons for social workers. For over a decade, scholars have argued for the applicability of CRT to social work pedagogy (Kolivoski et al., 2014; Razack & Jeffery, 2002). CRT provides a robust theoretical framework for anti-racism pedagogy by focusing on the interrogation and analysis of institutional racism, power relations, justice, and equity in teaching.

We present pedagogical examples highlighting the tenets, differential racialization, counterstorytelling, and intersectionality (Bell, 1995; Crenshaw et al., 1995). Counter-storytelling, as a CRT tenet, helps to decenter whiteness and highlight the contributions of BIPOC scholars, organizations, and institutions to U.S. history, the field of social work, and social justice achievements and struggles. For example, an explicit anti-racism pedagogy requires the acknowledgment and embrace of the contributions of Black social work pioneers whose perspective of social work centered an intimate understanding of the strengths and needs of the Black community and the commitment to advance social justice (Bent-Goodley et al., 2017). Embracing counter-narratives amplifies the voices of BIPOC social workers and communities. This counter-dominant approach to pedagogy works to center the empowerment and liberation of BIPOC and other marginalized students (Kailin, 2002). In addition, CRT's tenets of differential racialization and intersectionality provide a lens for social work education to expose the saliency of race to interlocking oppressions and the discourse of racial justice and equity in teaching. CRT's action orientation also answers the urgency of the CSWE Task Force on Anti-racism's call to move social work education from analysis to anti-racism action.

Critique of social work practice with children, youth, and families

An initial step in advancing anti-racism in social work education is acknowledging social work's complicity in harmful and racist policies and practices. Our profession's history with children, youth, and families is fraught by a legacy of enduring racism in two ways: First, through explicit support and practice enforcement of social policies specifically designed to oppress and marginalize BIPOC (Ioakimidis & Trimikliniotis, 2020; Marcynyszyn et al., 2012); second, through adopting the conceptualization of the heterosexual biological, nuclear family resulting in the economic and social punitive effects on BIPOC families (Gerstel, 2011; Peterson, 2013; Trattner, 1999). The following section interrogates social work's role in racially biased child welfare and school social work practices.

Since its inception, social work's entanglement in harmful policies to nonwhite children is well documented. Family separations of enslaved persons were likely a source of inspiration for the doctrine of *"Manifest Destiny,"* which propelled the removal of Indigenous people who stood in the way of westward expansion (Miller & Miller, 2006). Thus, a *"Kill the Indian, Save the Man"* policy found its home in establishing the Indian boarding school system, whose primary purpose was to annihilate native cultures through forced assimilation. In this context, social workers played a key role in the process of children's removal and separation from their families (Yellow Bird & Chenault, 1999).

Similarly, social work's anti-Black racist practices were manifest in the exclusion of social services in social casework and settlement houses to Black families and the profession's general acceptance of segregation in schools. The 1965 Moynihan Report became a central public discourse that identified and targeted Black families as "welfare" recipient families whose stereotype apex was President Reagan's "welfare queen." From the friendly visitors onward, social workers have consistently investigated which family members live in the home. Black families were particularly scapegoated and became targets for subsequent "welfare reform" (Levenstein, 2000). Social work has failed to address the institutional racism that promoted these stereotypes in child welfare services, which played a role in the early 1980's war on drugs, subsequent child protection, juvenile detention, and school to prison pipeline (Alexander, 2012; Roberts, 2014).

Black children are 13.71% of the population yet account for 22.75% of children in foster care. Indigenous children account for 1% of the population yet disproportionately comprise 2.4% of children in foster care (Annie E. Casey Foundations, 2018). Kolivoski et al. (2014) assessed this overrepresentation of BIPOC children as a product of racial bias; the direct results of these practices have devastating impacts on BIPOC children, and social work cannot deny its complicity. Calero et al. (2017) estimates between 14–80% of children placed into foster care face early criminalization and end up in the foster care-to-prison pipeline.

In addition, recent family separations at the United States' southern border constitute the latest racist and xenophobic travesties visited upon Latinx children who continue to be traumatized and held without family contact. These border camps have been continuously scrutinized for neglecting the public health of those detained and placing them at increased risk for contracting COVID-19. Social work's absence of leadership and depoliticized stance (Maylea, 2021) is conspicuous considering the atrocities being perpetrated.

Early school social work practice began between 1906 and 1907, serving public schools with an increase of African Americans moving from the south to the north, immigrants, working-class, and poor children in schools (Bye & Alvarez, 2006; Tyack, 1974, cited in K. L. Phillipo & Blosser, 2013).

School social workers face a significant amount of pressure when functioning as a part of a system that has been complicit over time with unjustly labeling BIPOC and low socioeconomic groups with academic and behavioral problems resulting from institutional discrimination (K. Phillipo & Stone, 2011). For example, during the pandemic, disparities in access to technology and reliable internet access disproportionately affected BIPOC families and students living in poverty. Vaughans and Spielberg (2014) emphasized the importance of systemic change by understanding the history of oppression for BIPOC groups; the pandemic demonstrates that social work must also address contemporary manifestations (e.g., anti-racism action in some locations meant getting children hotspots).

CRT a framework for anti-racism pedagogical examples

The critique discussed above demonstrates the importance of social work educators' incorporating anti-racism pedagogy into their teaching practice. To date, social work education's content on children, youth, and families has mainly focused on the psychological and behavioral dynamics of trauma, abuse, and neglect without critical examination of how racism structures the developmental context. Moreover, the social work curriculum does not adequately teach students how white supremacy has socially constructed views of BIPOC children, youth, and families. Nor does mainstream social work education build students' capacity to critically analyze and dismantle policy, practices, and structures that uphold white supremacy or provide tools to build networks of solidarity inside and outside the profession to create policy, practices, and systems grounded in anti-racism.

The dual pandemics of racism and COVID-19 compelled the work of the Anti-Racism Task Force subcommittee to provide examples of how CRT approaches can be incorporated as a case for learning about anti-racism. The following section offers anti-racism pedagogical examples that illustrate strategies to employ an anti-racism lens to historical and current societal issues within elective and core social work courses. We start with a module using the CRT tenet of differential racialization to teach about family separation. Our second example demonstrates the integration of CRT into a school social work course. Next, we use CRT to examine the social construction of immigration as a racialized issue. Our final example illustrates the impact of counter-storytelling through a module addressing racial disproportionality in the Texas Child Welfare system. These anti-racism pedagogical examples critically interrogate the role of social work in perpetuating existing paradigms and practices and offer opportunities to change social work's role.

Teaching about family separation using a differential racialization lens

The lesson on differential racialization described below occurs mid-way through a 15-week MSW course. The purpose of this lesson is to facilitate students' ability to examine how the historical legacy of structural racism continues to this day.

Differential racialization suggests that different ethnic-racial groups – such as BIPOC – have been racialized (socially constructed and treated) in different ways throughout history according to the needs of the dominant group (Crenshaw et al., 1995). For example, COVID-19 fears paired with scapegoating tactics employed by the Trump administration have instigated a resurgence of violence against Asian Americans, despite longstanding depictions of the model minority myth. This rise in anti-Asian racism is reminiscent of historical "yellow peril" narratives and imagery (Li & Nicholson, 2021).

This activity involves students in co-creating a timeline of U.S. racial history. In a class of 21, students work in five groups of 4–5 students to conduct "research" and create "timeline entries" for their assigned racialized group (i.e., Black, Indigenous, Latinx, Asian, Middle Eastern/Arab). Using course readings, internet search engines, and personal/familial experience for their assigned racialized group, students create timeline entries of historical events, policies, or social milestones fueled by racism covering the entirety of U.S. history – from its colonial history to the present time – using a set of Post-it Notes. The instructor designs a "timeline" graphic to exhibit the student entries. Students take a 5-minute "gallery walk" of the timeline once it is completed and debrief their reactions.

The second, most essential, part of the activity examines family separation using a differential racialization lens. The instructor selects students' timeline entries that correspond with chattel slavery, "Indian" boarding schools, the War on Drugs, and the founding of ICE to practice using differential racialization as an analytic tool. Students discuss the following questions: 1) Who is the target group, and how were they socially constructed at the time? 2) What policy, institutional practices, or social dynamics were reinforced or informed by the social construction of this group? 3) What were the consequences for the families and communities affected? Collectively, the class identifies how racism contributed to family separation. Students give examples of current events depicting "children in cages" being detained at the Southern border. Some students connect personally to their families' fear of immigration raids and deportations. If necessary, the instructor connects the establishment of ICE with the "War on Terror" rhetoric that racialized Muslims as terrorists after 9/11. The discussion explores how the trade of enslaved people strategically separated children from their parents. The instructor also helps students articulate how boarding schools coercively removed Indigenous children from their homes aside from enforcing assimilation.

Students often need assistance to critically examine how the War on Drugs may be another example of family separation. The instructor helps connect students' remarks about the pervasiveness of mass incarceration in the U.S. and its racial bias toward Black and Latino men to family separation. If necessary, the instructor shares the recent rise in the over-incarceration of women of color. Students can better articulate how the mass incarceration of BIPOC ultimately results in family separation by unpacking gendered experiences of criminalization. The instructor ends by summarizing how "family separation" is both a tool and a result of white supremacy and underscores the utility of the differential racialization tenet in examining how structural racism manifests differently across social groups, time, and contexts.

Teaching school social work practice integrating critical race theory

In an elective 14-week course, students learn the historical and current working practices with children and families in public school settings. During the dual pandemics, school social workers were on the front lines as essential workers helping families meet basic needs and supporting students in virtual learning while observing Black men, women, and children being murdered. This "new normal" requires school social workers to be equipped with renewed strategies to lead anti-racism and social justice efforts within the school and community settings. CRT provides an entry place to discuss anti-racism and works to challenge institutionalized racism. The examination of intersectionality invites assessment and engagement on multiple levels. The ecological and data-informed practices offer opportunities for counter-narratives.

In the assignment, students review the state General Statutes regarding suspension and expulsion. They examine their field placement school district's policies and the report on suspension and expulsion of students. Students interrogate the data, seeking the common circumstances resulting in student suspensions. Students report on the variance of the data among gender, age, and race. Students are challenged to develop and discuss hypotheses of contributing factors to the variances. They research alternatives to suspension in the school system and evaluate the effectiveness of the options in reducing suspension rates, including data or observations and reports from school staff. The final element of the assignment includes advocacy with sound recommendations for the school administration and board that would improve the current suspension policies and procedures.

The assignment and process are intended to illuminate the disparities in discipline and sound the alarm for improved policies and practices to increase children's ability to graduate instead of entering the juvenile justice system and the school-to-prison pipeline. Social workers and educators must seek organizational change to address the structural inequities. These changes include equipping school staff with knowledge and context of the unique issues that BIPOC families face to guide how schools effectively support students' academic success.

Using CRT to examine the discourse of immigration as a racialized issue

In an elective eight-week MSW course, CRT is used to discuss race as a forced social construct on immigrants and refugees in the U.S. In the course, discussions cover immigration policies in the U.S. from the end of World War I when the anti-immigrant sentiment started to grow to the present time with broad executive orders and actions targeting immigrants to discuss how the desirability of different groups has changed based on the dominant discourse. The class lectures and discussions focus on top origin countries of immigrants and refugees and specific populations of concern, including older adult immigrants, mixed-status families, queer immigrants and refugees, unaccompanied and separated children, and survivors of human trafficking. Our discussions highlight the centrality and intersectionality of racism.

As part of this course, students work individually or in groups to complete two assignments and a quiz. For the first assignment, "critical reflection on immigration policies," students deliver a PechaKucha-style presentation about a specific group and discuss how immigration policies have impacted this group since 1925. By watching the presentations,

students see that while anti-immigration sentiment has grown over the years, governing policies impacted diverse groups of immigrants and refugees differently. In this context, after the presentations, students write a reflective discussion post about the role of race as a social construct governed by immigration law in the U.S.

For the second assignment, "taking a step," students focus on a current oppressive immigration policy and write a letter or social media post addressed to the President (or local government officials if applicable), critiquing the policy and offering alternative solutions. This year students discussed the Trump administration's weaponization of COVID-19 under U.S. health law, Title 42, allowing mass deportation of asylum seekers at the U.S.–Mexico border. Students criticized the current administration for extending the policy while other immigrants and refugees entered the country through the southern border or by flights and in the absence of convincing evidence that this policy will prevent spread of COVID-19. The alternative solutions discussed by students were providing access to COVID-19 vaccination, fair opportunity to seek asylum for every human being, and addressing the legality of using Title 42.

Students take a scavenger hunt style quiz for the last activity, reading through a series of hints to answer questions and write short critical reflection posts about working with immigrant and refugee children, youth, and families. In this activity and class lectures, students are encouraged to acknowledge the intersectionality and fluidity of immigrants' and refugees' identities and the centrality and intersectionality of racism. This course prepares social work students to work with a growing population of minoritized immigrants and refugees in the U.S.

CRT in action: addressing disproportionality in Texas child welfare, the Texas model

This application of anti-racism pedagogy in an MSW community practice course applies the CRT tenets of counter-storytelling and race and racism as endemic in society. The course hosted Joyce James, a Black social work pioneer. Joyce James developed the Texas Model to address disproportionality in Texas Child Welfare, an anti-racism policy-practice model grounded in community voice, solidarity, and accountability to improve outcomes for children and families. This example illustrates anti-racism as an "active process of identifying and eliminating racism by changing systems, organizational structures, policies and practices and attitudes" (NAC, n.d.) through the collaborative building of solidarity networks (CSWE, 2021).

In 2004, Joyce became Assistant Commissioner for Child Protective Services; she joined the state's collaboration with Casey Family Programs. The resulting Texas Model uses a cross-systems approach linking child welfare, juvenile justice, education, health and mental health, workforce, and other systems to address disproportionality statewide (James et al., 2008). The model encompasses lessons from Joyce's earlier work in developing Project HOPE (Helping Our People Excel) in collaboration with parents, youth, and allies from local churches, nonprofits, county, state, and federal agencies dedicated to health, welfare, and protection of children. Within the Texas Model, understanding and analyzing institutionalized racism and building solidarity to undo racism is a primary concern. Through the Undoing Racism workshop provided by the People's Institute for Survival and Beyond, CPS leadership and community partners learned to; 1) analyze power, 2) define racism, 3) identify manifestations of racism, 4) learn from history, 5) share culture, and 6) organize to undo racism within systems, institutions and at the community level (The People's Institute for Survival and Beyond, 2007 as cited in James et al., 2008).

Community engagement strategies are used to enroll community members and build local allies by making the problem of disproportionality visible by sharing real numbers and stories through constituents' voices. These strategies then move to *community leadership* which expands leadership to the appropriate community level and offers Undoing Racism training to reinforce the committee members as agents of social change. *Community organization* is grounded in the assumption that the community must lead the work in partnership with CPS. *Community accountability* is working toward desired outcomes and measurable results guided by the belief that communities are the "owners" of the solutions to achieve sustainable safety, permanency, and well-being for their children (Seymour, 2007).

Her core message to social workers resonates with the crises of the dual pandemics:

> Without an analysis and understanding of institutionalized racism, we do harm. As all of our helping systems produce disparate outcomes that have a collective impact, there is a critical need for us to understand how all these systems create this harm. Once we have that analysis and understanding, we must become critical lovers of our systems to make change, and we must invite in the community to inform us of how systems can better serve us (J. James, personal communication, June 2, 2021). The impact of COVID on communities of color reflects the ongoing impact of institutional and structural racism on poor communities of color. It is not surprising that though all populations are susceptible to the virus, the conditions that have long existed for communities of color have become even more visible with COVID and the resulting continued oppression and community loss that emanates from the long history of racism and the collective impact of systems that come every day in the name of help. Yet, the data across all helping systems tell a very different story . . . the lack of a clear analysis of racism and the inherent nature of it in the distribution or absence of resources leading to the unnecessary deaths of Black and Brown people. This is not a new phenomenon! This is not a new experience for a different set of people. It is the same experience that the same people have lived with since the inception of this county (J. Joyce, personal communication, Oct 12, 2021).

After the presentation, students break into small groups to reflect on the communities their internships are in and discuss the following questions: 1) What are the organic strengths and resources within the community that your agency and other partners could support? 2) How do institutionalized racism and racial disparities across systems impact the ability of families and communities to protect and care for their children? 3) What are the action steps you can take to embrace anti-racism and become a critical lover of your internship?

Discussion

In sum, the dual pandemics triggered the call for a greater emphasis and sustained awakening to advance anti-racism in social work education and practice with BIPOC children, youth, and families. Hillis et al. (2021) urges a holistic response, in congruence with the social work sphere of practice, policy, and support of children, youth, and families to include direct financial assistance, health coverage, mental health treatment, and enhanced resources to the foster care system. As we educate new social workers, our hope for anti-racism social work practice begins with anti-racism pedagogy. From the inequitable distribution of benefits, the disproportionate number of BIPOC children in the child welfare system, separating BIPOC children from their families to the erasure of BIPOC voices in the curriculum, social work has an explicit hand in the far-reaching consequences of racism. The current pushback against CRT by several states across the U.S. demonstrates that unseating white supremacy will not be easy (Abrams & Detlaff, 2021). Social work's future depends on CSWE intentionally moving social work education beyond momentary pledges that carry an illusion of being on the right side of racial justice and embracing the hard work of active anti-racism in social work education.

Implications for social work education and practice

Forming the CSWE Task Force on Anti-racism was a much-needed first step for anti-racism within social work education. However, as evidenced in our critique and pedagogical examples, small incremental steps are insufficient given social work's long history of complicity in racism. This paper echoes the suggestions of the Task Force for Anti-racism. Our recommendations for anti-racism pedagogy in social work include:

Atoning for social work's complicity in structural racism

Just as it is essential for individuals to challenge their racism and biases, we must challenge our profession's systemic racism for change to occur. Making amends for our profession's historical and contemporary legacy of racism involves being intentional about recognizing and repairing past social work values, practices, and policies that upheld white supremacy or reinforced the racialization and oppression of vulnerable families and communities

Centering the policy-practice models used by BIPOC practitioners

In social work's attempt to reckon with its complicity in racism, it is essential to center the contributions of BIPOC social work pioneers who provided services supporting diverse communities. To counter the dominant deficit-narrative, stop epistemic erasure, and move social work toward its anti-racism future, we must highlight the counter-narratives of BIPOC practitioners that have been pushed to the side. Social work educators can integrate this knowledge into class content through formal scholarly work, podcasts, music, blogs, art, poetry, comedy, guest speakers, and by incorporating assignments and assessments that include collaborative projects, oral presentations, blogs, podcasts, art, and the creation of media.

Building social workers' capacity for critical race praxis

Social work pedagogy should develop the knowledge and skills necessary to put theory into practice (praxis) in ways that recognize, critically examine, and address the structural racism that creates social inequity. CRT can be one of many theoretical frameworks used to disrupt white supremacy in social work education. This intentional anti-racism action will transform social work education and create an impetus that changes social work practice to processes that oppose and challenge systemic and structural inequities.

Infusing anti-racism pedagogy throughout the curriculum

Social work education must move to anti-racism pedagogy throughout the curriculum to unpack its complicity in white supremacy and structural racism. Integrating anti-racism pedagogy in every class, not just "diversity" courses, is essential to advancing anti-racism in social work. Critical analysis of race and racism enhances curricular content and instruction irrespective of the course topic, as illustrated in the pedagogical examples offered in this paper. For instance, in developing this manuscript, one of the authors reevaluated the design of her course on critical refugee and migration studies to incorporate CRT.

Examining our institutional practices and policies

Anti-racism pedagogy requires social transformation inside and outside of the classroom. To effectively do this work, we must be willing to invite alumni, students, field advisors, and other constituents to share their experiences of racism within our programs and universities and join with us to build accountable institutions grounded in anti-racism. We need to develop intentional networks of solidarity within our universities and the profession, and our communities.

Conclusion

The history of social work's complicity in white supremacy and systemic racism is hard to overlook, yet social work has failed to take anti-racism action that unseats white supremacy (Beck, 2019). Social work's efforts fall far short of the Code of Ethics that it claims to embrace. Our call for anti-racism in social work joins decades of calls for anti-racism. This discussion offers social work a pathway to promote racial justice and dismantle systemic racism and white supremacy within social work practice. The dual pandemic highlighted the urgency; let the call for a sustained awakening of advancing anti-racism in social work education and practice be heard. It is no longer enough for CSWE and social workers to "claim" not to be racist – we must lead in anti-racism.

Disclosure statement

No potential conflict of interest was reported by the author(s).

References

Abrams, L., & Detlaff, A. (2021). Why social work needs to double down on critical race theory. https://labramsucla. medium.com/why-social-work-needs-to-double-down-on-critical-race-theory-4322296754b4

Aldana, A., & Vazquez, N. (2020). From colour-blind racism to critical race theory: The road towards anti-racist social work in the United States. In G. Singh & S. Masocha (Eds.), *Anti-racist social work: International perspectives* (pp. 129–148). Red Globe Press.

Alexander, M. (2012). *The new Jim Crow: Mass incarceration in the age of colorblindness.* The New Press.

Annie E. Casey Foundations. (2018). *2018 kids count data book: State trends in child well-being.* www.aecf.org/databook

Beck, E. (2019). Naming white supremacy in the social work curriculum. *Affilia: Journal of Women and Social Work, 34* (3), 393–398. https://doi.org/10.1177/0886109919837918

Bell, D. (1995). Who's afraid of critical race theory? *University of Illinois Law Review, 1995*(4), 893–910. https:// heinonline.org/HOL/Page?handle=hein.journals/unilllr1995&div=40&g_sent=1&casa_token=&collection=journals

Bent-Goodley, T., Snell, C. L., & Carlton-LaNey, I. (2017). Black perspectives in social work practice. *Journal of Human Behavior in the Social Environment, 27*(1–2), 27–35. https://doi.org/10.1080/10911359.2016.1252604

Blakeney, A. M. (2005). Anti-racist pedagogy: Definition, theory, and professional development. *Journal of Curriculum & Pedagogy, 2*(1), 119–132. https://doi.org/10.1080/15505170.2005.10411532

Brice, T. S., & McLane-Davison, D. (2020). The strength of Black families. In A. N. Mendenhall & M. M. Carney (Eds.), *Rooted in strengths: Celebrating the strengths perspective in social work* (pp. 25–37). Jayhawk Ink.

Bye, L., & Alvarez, M. (2006). *School social work: Theory to practice.* Thompson/Brooks/Cole.

Calero, S., Kopić, K., Lee, A., Nuevelle, T., Spanjaard, M., & Williams, T. (2017). On the problematization and criminalization of children and young adults with non-apparent disabilities. *The Ruderman White Paper.* https:// rudermanfoundation.org/white_papers/criminalization-of-children-with-non-apparent-disabilities/

Centers for Disease Control (CDC). (2021a). *COVID-19: People with certain medical conditions.* https://www.cdc.gov/ coronavirus/2019-ncov/need-extra-precautions/people-with-medical-conditions.html

Centers for Disease Control (CDC). (2021b). *Risk for COVID-19 infection, hospitalization, and death by race/ethnicity* http://www.cdc.gov/coronavirus/2019-ncov/covid-data/investigations-discovery/hospitalization-death-by-race-ethnicity.html

Corley, N. A., & Young, S. M. (2018). Is social work still racist? A content analysis of recent literature. *Social Work, 63*(4), 317–326. https://doi.org/10.1093/sw/swy042

Council on Social Work Education (CSWE). (2021). *An update from the anti-racism task force.* https://www.cswe.org/ News/General-News-Archives/An-Update-From-the-Anti-Racism-Task-Force

Crenshaw, K., Gotanda, N., Peller, G., & Thomas, K. (Eds.). (1995). *Critical race theory: The key writings that formed the movement.* The New Press.

Dominelli, L. (1997). *Anti-racist social work: A challenge for white practitioners and educators.* British Association of Social Workers (BASW) Practical Social Work.

Economic Policy Institute (EPI). (2020). *Black workers face two of the most lethal preexisting conditions for coronavirus—racism and economic inequality.* https://www.epi.org/publication/black-workers-covid/externalicon

Gerstel, N. (2011). Rethinking families and community: The color, class, and centrality of extended kin ties. *Sociological Forum, 26*(1), 1–20. https://doi.org/10.1111/j.1573-7861.2010.01222.x

Harty, J. (2021). *Black contributions to social welfare & social work history: A legacy of Black self-help, resistance and liberation.* Equity & Inclusion Speaker Series. http://www.bu.edu/ssw/files/2021/02/Harty-BlackSWSW-Boston-20210225.pdf

Hassouneh, D. (2006). Anti-racist pedagogy: Challenges faced by faculty of color in predominantly white schools of nursing. *Journal of Nursing Education, 45*(7), 255–262 https://doi.org/10.3928/01484834-20060701-04.

Hawkins, D. (2020). Differential occupational risk for COVID-19 and other infection exposure according to race and ethnicity. *American Journal of Industrial Medicine, 63*(9), 817–820. https://doi.org/10.1002/ajim.23145

Hillis, S., Blenkinsop, A., Villaveces, A., Annor, F., Liburd, L., & Massetti, G. (2021). COVID- 19 associated orphanhood and caregiver death in the United States. *Pediatrics, 148*(6), e2021053760. https://doi.org/10.1542/peds.2021-053760

Ioakimidis, V., & Trimikliniotis, N. (2020). Making Sense of Social Work's Troubled Past: Professional Identity, Collective Memory and the Quest for Historical Justice. *British Journal of Social Work, 50*(6), 1890–1908. https://doi.org/10.1093/bjsw/bcaa040

James, J., Green, D., Rodriguez, C., & Fong, R. (2008). Addressing disproportionality through undoing racism, leadership development, and community engagement. *Child Welfare, 87*(2), 279–296.

Kailin, J. (2002). *Anti-racist education: From theory to practice.* Rowman & Littlefield.

Karaye, I. M., & Horney, J. A. (2020). The impact of social vulnerability on COVID-19 in the U.S.: An analysis of spatially varying relationships. *American Journal of Preventive Medicine, 9*(3), 217–325. https://doi.org/10.1016/j.amepre.2020.06.006externalicon

Khazanchi, R., Evans, C. T., & Marcelin, J. R. (2020). Racism, not race, drives inequity across the COVID-19 continuum. *JAMA Network Open, 3*(9), e2019933. https://doi.org/10.1001/jamanetworkopen.2020.19933

Kishimoto, K. (2018). Anti-racist pedagogy: From faculty's self-reflection to organizing within and beyond the class-room. *Race Ethnicity and Education, 21*(4), 540–554. https://doi.org/10.1080/13613324.2016.1248824

Kolivoski, K. M., Weaver, A., & Constance-Huggins, M. (2014). Critical race theory: Opportunities for application in social work practice and policy. *Families in Society, 95*(4), 69–276. doi:10.1606/1044-3894.2014.95.36

Levenstein, L. (2000). From innocent children to unwanted migrants and unwed moms: Two chapters in the public discourse on welfare in the United States, 1960–1961. *Journal of Women's History, 11*(4), 10–33. https://doi.org/10.1353/jowh.2000.0009

Li, Y., & Nicholson, H. (2021). When "model minorities" become "yellow peril"—Othering and the racialization of Asian Americans in the COVID-19 pandemic. *Sociology Compass, 15*(2), e12849. https://doi.org/10.1111/soc4.12849

Marcynyszyn, L. A., Bear, P. S., Geary, E., Conti, R., Pecora, P. J., Day, P. A., & Wilson, S. T. (2012). Family group decision making (FGDM) with Lakota families in two tribal communities: Tools to facilitate FGDM implementation and evaluation. *Child Welfare, 91*(3), 113–134.

Maylea, C. (2021). The end of social work. *British Journal of Social Work, 51*(2), 772–789. https://doi.org/10.1093/bjsw/bcaa203

Miller, R. J., & Miller, R. (2006). *Native America discovered and conquered: Thomas Jefferson, Lewis & Clark, and manifest destiny.* Greenwood Publishing Group.

Millett, G. A., Jones, A. T., Benkeser, D., Baral, S., Mercer, L., Beyrer, C., Honermann, B., Lankiewicz, E., Mena, L., Crowley, J. S., Sherwood, J., & Sullivan, P. S. (2020). Assessing differential impacts of COVID-19 on black communities. *Annals of Epidemiology, 47*, 37–44. https://doi.org/10.1016/j.annepidem.2020.05.003

Moore, J., Ricaldi, J., Rose, C., Fuld, J., Parise, M., Kang, G., Driscoll, A., Norris, T., Wilson, N., Rainisch, G., Valverde, E., Beresovsky, V., Brune, C., Oussayef, N., Rose, D., Adams, L., Awel, S., Villa, J., Meaney-Delman, D., & Honein, M. (2020). Disparities in incidence of COVID-19 among underrepresented racial/ethnic groups in counties identified as hotspots during June 5 –18, 2020 — 22 States, February–June 2020. *MMWR, 69*(33), 1122–1126. http://dx.doi.org/10.15585/mmwr.mm6933e1

NAC (n.d.). NAC International Perspectives: Women and Global Solidarity, Antiracism. Resources for Racial Justice. https://libguides.usu.edu/racialjustice/concepts

Okoh, A., Sossou, C., Dangayach, N., Meledathu, S., Phillips, O., Raczek, C., Patti, M., Kang, N., Hirji, S., Cathcart, C., Engell, C., Cohen, M., Nagarakanti, S., Bishburg, E., & Grewal, H. (2020). Coronavirus disease 19 in minority populations of Newark, New Jersey. *International Journal for Equity in Health, 19*(93). https://doi.org/10.1186/s12939-020-01208-1

Peterson, C. (2013). The lies that bind: Heteronormative constructions of "family" in social work discourse. *Journal of Gay & Lesbian Social Services, 25*(4), 486–508. https://doi.org/10.1080/10538720.2013.829394

Phillipo, K. L., & Blosser, A. (2013). Specialty practice or interstitial practice? A recon- sideration of school social work's past and present. *Children & Schools, 35*(1), 19–31. https://doi.org/10.1093/cs/cds039

Phillipo, K., & Stone, S. (2011). Toward a broader view: A call to integrate knowledge about schools into school social work research. *Children & Schools*, *33*(2), 71–81. https://doi.org/10.1093/cs/33.2.71

Razack, N., & Jeffery, D. (2002). Critical race discourse and tenets for social work. *Canadian Social Work Review*, *19*(2), 257–271. http://www.jstor.org/stable/41669763

Roberts, D. E. (2014). Child protection as surveillance of African American families. *Journal of Social Welfare and Family Law*, *36*(4), 426–437. https://doi.org/10.1080/09649069.2014.967991

Rodriguez-Lonear, D., Barcelo, N. E., Akee, R., & Carroll, S. R. (2020). American Indian reservations and COVID-19: Correlates of early infection rates in the pandemic. *Journal of Public Health Management and Practice*, *24*(4), 371–377. https://doi.org/10.1097/PHH.0000000000001206externalicon

Seymour, J. (2007). *Engaging communities and taking a stand for children and families: Leadership development and strategic planning in the Texas child welfare system*. Casey Family Programs Texas State Strategy and Texas Child Protective Services.

Tobolowsky, F., Gonzales, E., Self, J., Rao, C., Keating, R., Marx, G., McMichael, T., Lukoff, M., Duchin, J., Huster, K., Rauch, J., McLendon, H., Hanson, M., Nichols, D., Pogosjans, S., Fagalde, M., Lenahan, J., Maier, E., Whitney, H., & Kay, M. (2020). COVID-19 outbreak among three affiliated homeless service sites — King County, Washington. *MMWR*, *69*(17), 523–526. http://dx.doi.org/10.15585/mmwr.mm6917e2

Trattner, W. I. (1999). *From poor law to welfare state: A history of social welfare in America* (6th ed.). The Free Press.

Tyack, D. B. (1974). *The One Best System: A History of American Urban Education*. Harvard University Press.

Vaughans, K. C., & Spielberg, W. (2014). *The psychology of Black boys and adolescents*. Praegar.

Wallace, M., Hagan, L., & Curran, K. G. (2020). COVID-19 in correctional and detention facilities – United States, February-April 2020. *MMWR*, *69*(19), 587–590. http://dx.doi.org/10.15585/mmwr.mm6919e1externalicon

Waltenburg, M. A., Victorroff, T., & Rose, C. E. (2020). Update: COVID-19 among workers in meat and poultry processing facilities – United States, April-May 2020. *MMWR*, *69*(27), 887–892. http://dx.doi.org/10.15585/mmwr.mm6927e2externalicon

Yellow Bird, M. J., & Chenault, V. (1999). *The role of social work in advancing the practice of indigenous education: Obstacles and promises in empowerment-oriented social work practice*. ERIC. https://eric.ed.gov/?id=ED427911

Advancing critical race pedagogical approaches in social work education

Patrina Duhaney, Liza Lorenzetti ⓘ, Kaltrina Kusari and Emily Han

ABSTRACT

The recent COVID-19 pandemic drew a sharp focus on existing inequities for racialized communities in Canada and globally. A paucity of research-informed transformative learning models in social work has resulted in the persistent centering of Western ways of knowing. Current efforts do not adequately address the nuances of systemic and structural racial inequities, leaving students unprepared to deal with these issues in the classroom and in practice. We propose critical race pedagogy as an essential framework to promote and enrich social work learning environments where students can engage in courageous conversations about race, racism, power and oppression.

Canada is often portrayed as a paradise of multicultural and racial equity, yet it is steeped in a history of colonialism, white supremacy, and systemic and institutional racism. Racism and anti-Blackness are inextricably linked (Cooper, 2006) to colonization, 200 years of legislated slavery (Cooper, 2006), and genocide against Indigenous peoples (Truth and Reconciliation Commission of Canada [TRC], 2015), which persist in various forms today (Goulet, 2018). Systemic anti-Blackness encompasses a myriad of intersecting forms of historically-embedded, institutionalized and state-sanctioned violence against Black communities. These include, but are not limited to, mass incarceration, the school-to-prison pipeline, the gentrification of low-income communities, the exposure of low-income areas to pollution and environmental hazards, and the traumatic impact of racism on Black people (Esposito & Romano, 2016). Black people are more likely to be arrested, have restrictive bail conditions, and receive longer sentencing than their white counterparts (Maynard, 2017). They also experience higher rates of abuse compared to the general prison population (Maynard, 2017). Police engage in disproportionate and unjustified use of violence against Black people, but are seldom held accountable for their actions (Ontario Human Rights Commission, 2018). Racism also continues "to underpin the colonization experiences of Indigenous people" (Goulet, 2018, p. 77), with a barrage of assaults that include "resource exploitation of Indigenous lands, residential school syndrome ... expropriation of lands, extinguishment of rights, wardship, and welfare dependency" (Alfred, 2009, p. 43).

The historic and intensified focus of racial violence on Black and Indigenous lives has become increasingly evident within global circles, resulting in international reprimand (United Nations, 2016). The current COVID-19 pandemic has also drawn a glaring spotlight on the relentless presence of racism in Canada and globally. Although historical and current conditions provide unequivocal evidence that racism has deleterious health and well-being outcomes for racialized and Indigenous communities, social work education has fallen short in preparing students for a critical commitment to anti-racist practice. Further, neoliberalism has extended the breeding ground for white supremacy, including the privatization of racial discourse (Coxshall, 2020). Given social work's historic and ongoing complicity in maintaining and reinforcing colonial legacies which negatively impact Black, Indigenous, and racialized communities, the social work profession has an obligation to redress these harms. In particular, it must contend with racial injustices within its profession, educational practices, policies and procedures. The social work profession experiences increasing pressure to engage with, learn and teach anti-racist theories, practice, and research (Suoranta & Moisio, 2006). However, approaches such as anti-oppressive practice (AOP) and structural social work are limited in their capacity to center both anti-racism and decolonization (Goulet, 2018; Parker & Stovall, 2004). Critical race theory, derived from legal and social science disciplines (Crenshaw et al., 1995), has been identified as a core theoretical framework that forefronts anti-racist approaches in social work. However, social work curriculum in Canada continues to omit critical pedagogies that center race, anti-racism and critical race theory. We propose critical race pedagogy as an essential framework to promote and enrich social work learning environments. From this pedagogical approach, students and educators can engage in high-level critical self-reflection and dialogue to disrupt the status quo, challenge dominant discourses, and advance social justice. In this article, we contend that by forefronting critical race pedagogies, students can engage in courageous conversations about race, racism, power and oppression to cultivate their agency and critical consciousness through ongoing reflection.

We position ourselves as critical anti-racist feminist scholars who are actively engaged in anti-racism work through teaching, research, scholarship, and activism. It is through these collective experiences that we conceptualize our paper. The lead author is a Black critical race scholar and social work activist with practice and research in the areas of anti-Black racism, critical race scholarship, social work pedagogy and the intersections between victimization and criminalization. The second author is a white Italian settler and community activist-scholar committed to anti-racism and decolonial practice. The third author is a white Albanian settler from Kosovo who examines the impact of colonial practices on the intersection of migration status, gender, and class for migrant populations in Canada and Kosovo. The fourth author is a second-generation Chinese immigrant, settler and a social worker with research and practice interests in family and intimate partner violence and its connections to institutionalized forms of violence.

Racial disparities during COVID-19: A dilemma for social work

The global pandemic drew a sharp focus on existing inequities for racialized communities at local, regional, and international levels. Although Black people represent 9% of the population in Canada, they account for 21% of COVID-19 cases (Allen, 2020). Research from the U.S. underscores the differential impacts of the disease on Black people, who constitute 13% of the population, but account for 30% of COVID-19 cases (Laurencin & Walker, 2020). A stark reality is that Black people are dying at a rate almost four times higher than the national average (Louis-Jean et al., 2020).

Many people became unemployed due to the economic consequences of the pandemic (Shadmi et al., 2020). Emerging data indicates that there are economic disparities for Black people who have been disproportionately affected by COVID-19 (Jedwab, 2020; Shadmi et al., 2020). Prior to the pandemic, approximately one third of Black Canadians had precarious employment, with even higher proportions among racialized, and newcomer women (Noack & Vosko, 2011). Approximately 61% of Black Canadians have seen their income decrease and 50% now struggle to meet their financial needs (Jedwab, 2020).

Amidst highly publicized incidents of police brutality, anti-Blackness, anti-Asian racism and xenophobia, unprecedented pressures on marginalized Canadians were punctuated during the summer of 2021 by the finding of unmarked graves of children victimized by Indian Residential Schools (Lindeman, 2021) and the hate-based murder of a Muslim family (Lupton & Dubinski, 2021). Additionally, the rise in COVID-related racist hate crimes against East and Southeast Asians is connected to the construction of racist discourses that depict Asians as a racial contagion (Mallapragada, 2021). According to Hager (2020), hate crimes against people of East Asian descent in Vancouver doubled in April 2020. The targeting of racialized communities is part of a historical legacy of racializing infectious diseases. Orientalist narratives of the "Yellow Peril" mark Asians as perpetual foreign threats within the nation (Mallapragada, 2021).

Racialized women in particular have experienced the dual crises of gender-based violence and systemic racism against the backdrop of a global pandemic. Safety measures that encourage people to stay at home, combined with the stressors of socioeconomic instability, continue to exacerbate the risk of domestic violence for women while creating barriers for accessing reliable support (Josephine & Hyman, 2021). This differential impact was highlighted in a study by Hyman and Vissandjee (2020), where 11.7% of immigrant women reported being extremely concerned about experiencing violence in the home during the pandemic, compared to 7% of Canadian-born women.

Racialized people, particularly Black and Indigenous communities, are vastly overrepresented among incarcerated populations in Canada (Maynard, 2017). The dual pandemic of systemic racism and COVID-19 (Thomas Bernard, 2020) underscore the multiple outbreaks observed in prisons across Canada. The close, enclosed quarters, high prevalence of chronic disease among inmate populations, and lack of adequate medical facilities in prisons increase racialized peoples' vulnerability to COVID (Shadmi et al., 2020). Alarmingly, during the pandemic, provinces like Ontario have seen expansions of policing powers, which disproportionately impact Black and Indigenous people who are often targets of police violence and surveillance (Khare et al., 2020).

The recent pandemic has further exposed historically-ladened professional and ethical challenges (Banks et al., 2020) in social work. As a profession, social work has historically played an active role in upholding whiteness and white supremacy, even in recent times, where the profession is under greater pressure to fight against racial violence and injustices. Dominelli (2002) notes that social work has been involved in upholding an oppressive status quo due its position as a unifying force within the nation-state's project of modernity, where social workers have attempted to unite and homogenize diverse groups through realization of citizenship and its entitlements. Yet social work has also been part of social control arrangements with inclusionary and exclusionary functions. Fortier and Wong (2018) trace the genealogy of social work in Canada and its history in colonial violence by connecting its professional roots to the work of missionaries and agents of the colonial state. Social workers have been complicit in the mass removal of Indigenous children through the residential school system and remain accountable, considering the disproportionate apprehension of Indigenous (TRC, 2015) and Black children today in child welfare services (Clarke, 2011). While some social workers have joined the struggle against racial injustices, the minor aid offered to marginalized groups through social workers' interventions work to pacify populations and relieve the settler state of its responsibility to enact systemic change (Fortier & Wong, 2018). Despite these challenges, social workers are uniquely positioned to identify and address the needs of racialized communities during the pandemic, however, they have faced numerous barriers. For example, a recent international study by Banks et al. (2020) surveying 607 participants found that social workers around the world faced ethical challenges during the COVID-19 pandemic due to limited resources for marginalized communities. Social workers faced ethical challenges balancing governmental and organizational restrictions with the needs of service users. For example, some workers decided to conduct face-to-face meetings with vulnerable clients despite social distancing policies, transported foster children in their own cars, or purchased COVID safety equipment for clients out of their own pockets (Banks et al., 2020).

Transforming social work education: Centralizing race and anti-racism

Heightened inequities and entrenched systemic racism refute claims of race neutrality within the social profession. Critical approaches to social work have become increasingly centralized by activist scholars situated in schools of social work, unraveling the tensions between the profession's comfort with the neo-liberal charity model and its claim as a social justice profession (Duhaney & El-Lahib, 2021). Critical social work encompasses a cluster of theories, such as feminist thought, structural approaches, antiracism, and anti-oppressive practice (AOP). Critical social work, with an overarching goal of social change and justice, links the personal with the political. It connects individual "clients" to the broader socio-political context they are situated within, with the goal of raising critical consciousness so that clients can resist oppressive social structures (Barak, 2019; Freire, 1970). Within the classroom, educators working from this perspective, engage students with participatory forms of instruction to guide them to confront prejudicial attitudes and commit to social justice (Estrada & Matthews, 2016). However, critical social work has been criticized for its preoccupation with analysis and critique that don't always articulate alternatives for actively changing the status quo (Williams, 2019).

Anti-oppressive practice (AOP) and structural social work (Mullaly, 1997) have been widely employed within social work classrooms as approaches to address racism, intersecting forms of oppression and structural inequalities. AOP draws upon multiple theories such as structuralism, feminism, and anti-racism, bringing together these distinct analyses under one umbrella (Baines, 2011; Brown, 2012). Social workers engaged in AOP are committed to consciousness-raising, not only on the individual or interpersonal level, but at the community and societal levels. Social workers reject the privileging of their "expert" knowledge at the cost of invalidating the valuable knowledge of clients, and work to initiate power-sharing processes with the client (Dominelli, 2002). Despite its ambitious goals of social change, AOP has been criticized for its lack of critical analysis of the state and its role in social work (Brown, 2012). Among its limitations is the fact that AOP rarely centers race, and anti-Blackness in particular, which is easily obfuscated under the broad umbrella of oppression.

Structural social work is committed to challenging and changing oppressive social structures. In particular, it critiques existing oppressive social, economic, and political institutions. It views social issues as an inherent, built-in part of our present social order, where social institutions function to oppress marginalized groups along the lines of class, gender, race, sexuality, disability, and other axes (Mullaly, 1997). Structural approaches are not concerned with centering or prioritizing one form of oppression over another; rather, structuralists understand intersecting forms of oppression as working together to create a total matrix of oppression. While structural social work and AOP are beneficial in exposing the undercurrents and systems that reproduce oppression, they are limited in centralizing race and anti-racist practice in social work.

Anti-racism in social work

Anti-racism is a political ideology from which to fight racism and respond to social and historic issues of slavery, colonialism, imperialism, and white supremacy. According to Dei (2000) anti-racism is an "an action-oriented educational strategy for institutional, systemic change to address racism and interlocking systems of social oppression. It is a critical discourse of race and racism in society that challenges the continuance of racializing social groups for differential and unequal treatment" (p. 27). Keating (2000) describes anti-racism as an activity that seeks to dismantle institutionalized racism and confront racist ideology to eliminate racism at the systemic and individual level. Antiracist ideology develops critical consciousness to understand and express the impact and experience of racism (Blakeney, 2005; Dei, 2000).

Emerging from social movements that aim to confront and dismantle racism, anti-racist pedagogy's goal is to organize for social transformation, not mere reform (Kishimoto, 2018). Antiracist pedagogy is described as a paradigm to understand and analyze the historical processes and constructs that produce systemic racism and uses praxis to transform oppressive social relations (Blakeney, 2005). It is a tool for analysis that identifies the social processes, structures and operations of society related to race and racial oppression. Anti-racist pedagogical approaches challenge existing structures of teaching and learning. As an instructional model, anti-racist pedagogy, focuses on developing students' capacities to address racism and interrogating whiteness and white privilege (Estrada & Matthews, 2016). Kishimoto (2018) notes that anti-racist pedagogy is not simply integrating racial content into curricula, but it encompasses *how* one teaches, even in courses where race is not the principal topic. As such, Kishimoto (2018) describes an anti-racist teaching approach as one that nurtures critical thinking skills and challenges students' assumptions; develops awareness of social positionality; decenters authority in the classroom and empowers students in their own learning; and creates a classroom community for collaborative learning.

Anti-racist perspectives have been criticized as reductionist of in-group ethnic differences. Single issue positioning has been problematized by the contention that individuals experience complex patterns of oppression at the intersections of race, class, gender, age, and other social positionalities. Single standpoint politics have also been criticized as creating hierarchies of oppression that become ultimately unproductive for achieving social justice goals (Williams, 1999). Indeed, a common criticism is the perception that anti-racism has been prioritized over other areas of oppression, although there are those who advocate for an integrated anti-racist approach that emphasizes a non-hierarchal model (Butler et al., 2003). However, Williams (1999) counters that collective categories within anti-racist perspectives allow for a sense of commonality around which to organize; this has often been the backbone of emancipatory social justice movements, whether in terms of class, race, ability, or gender.

Critical race theory and pedagogy: Centralizing race and systemic racism

Critical race theory emerged as an interdisciplinary movement and theoretical framework to explicate and resist racism. Like anti-racism, critical race theory is distinct in its specific focus on race. Critical race theory (CRT) is an interdisciplinary approach that is informed by critical legal studies, Black feminist theory, critical theory, feminism, liberalism, Marxism/neo-Marxism, poststructuralism, post-modernism, and neo-pragmatism (Crenshaw et al., 1995). CRT is premised on the belief that race is a social construct that is deeply embedded in society and permeates all aspects of social life (Ortiz & Jani, 2010). CRT can be used to address some of the shortcomings of critical pedagogy by centering discussions around race and other marginalized identities (Parker & Stovall, 2004). CRT creates a space where professionals and educators can deeply engage in discourse of race, inequality, privilege, power and oppression (Abrams & Moio, 2009; Coxshall, 2020).

Critical race pedagogy (CRP) emerged from African American practitioners and scholars producing scholarship rooted in CRT and/or Afrocentricity. CRP is "an analysis of racial, ethnic, and gender subordination in education that relies mostly upon the perceptions, experiences and counter-hegemonic practices of educators of color" (Lynn, 2004, p. 154). Building on the foundations of critical pedagogy, CRP forefronts race consciousness and empowers students to challenge notions of colorblindness, meritocracy, deficit thinking, linguicism, and other forms of subordination (Ledesma & Calderón, 2015). According to McCoy and Rodricks (2015), "critical race theory challenges us to move towards a constant unlearning and relearning that facilitates a practice of critique" (p. 55). Placing race at the center of analysis allows us to interrogate dominant discourses and develop strategies to eradicate the various forms of inequities and social disparities deeply woven in systemic and institutional structures.

Key principles of critical race pedagogy maintain that racism is endemic and deeply embedded in society; power is manifested in various forms; intersecting and overlapping identities inform racialized people's experiences; the voices and perspectives of racialized people have merit; and pedagogical approaches are both liberatory and transformative. Racism is endemic in society and has been shaped by institutions (Lynn, 1999). A critical race pedagogy emphasizes the need to recognize and challenge the assumption that the educational system is race neutral (Suoranta & Moisio, 2006). Another key aspect of critical race pedagogy is interrogating the ways in which various forms of knowledge are privileged while others are relegated to the margins (Nakaoka & Ortiz, 2018; Parker & Stovall, 2004). Critical race pedagogy centers the voices and lived experiences of racialized people and compels students to engage in a critical analysis of how power, oppression and systemic racism unfold in their lives. Critical race pedagogy validates the experiential knowledge of racialized students in order to deconstruct dominant ideologies in their classrooms (Alemán & Gaytán, 2017).

Rather than considering race as the only construct of importance, critical race pedagogy investigates the intersections of race, class, and gender, among other positionalities/axes of oppression (Lynn, 1999). CRT "offers a lens extensively exploring the intricacy of race and recognizing the concept of intersectionality, for better understanding inequality and oppression" (Campbell, 2014, p. 25). By critiquing the construction of race, educators can better understand the meaning of "whiteness," while promoting an intersectional approach that analyzes axes of oppression (Coxshall, 2020; Ledesma & Calderón, 2015).

Storytelling, a core tenet of CRP centers the voices of racialized people to inform nuanced understandings of their experiences of race and racism. Counter-storytelling can elevate epistemologies that counter "the metanarratives – the images, preconceptions, and myths – that have been propagated by the dominant culture of hegemonic whiteness as a way of maintaining racial inequality" (Treviño et al., 2008, p. 8).

CRT is recognized as both a liberatory teaching practice (Lynn, 2004) and a transformative tool for counteracting the devaluation of racialized students. This learning "expand[s] ... consciousness through the transformation of basic worldview and specific capacities of the self" (Elias et al., 1997, p. 3). This pedagogical approach provokes "students to question all taken-for-granted values, ideas, norms, and beliefs of experiences that comprise their dominant social paradigm" (Sagris, 2008, p. 1). Facilitating transformative pedagogical experiences is key to student learning and engagement as it encourages critical thinking and helps students develop confidence in decision-making. Liberatory approaches include – but are not limited to – critical pedagogy (Suoranta & Moisio, 2006), critical justice pedagogy (Sensoy & DiAngelo, 2014), anti-racist pedagogy (Harbin et al., 2019; O'Neill & Miller, 2015), decolonial pedagogy (Asher, 2009), feminist pedagogy (Bauer, 2000), Afro-centric pedagogy (Lynn, 2004), and engaged pedagogy (hooks, 1994). Critical thinking is a focal point of liberatory education and a cornerstone of conscious action (Freire, 1970). hooks (1994) liberatory approach to education is echoed within evolving critical antiracist feminist knowledge that accounts for intersectionality and intersecting oppression in social theory. Using liberatory pedagogy, students are engaged in critical thinking to promote and address oppression and injustices (Freire, 1970).

Employing critical race pedagogical approaches in the curriculum

There are myriad ways in which educators have used critical race pedagogy in the classroom, some of which include: (a) diversifying curriculum and course content; (b) encouraging student/educator reflexivity on intersecting identities; (c) exploring the implications of microaggressions; (d) making connections between everyday racism and larger systems of oppression; and (e) interrogating claims of race neutrality.

While enacting inclusivity through the curriculum is important, traditional modes of instruction often serve to exclude marginalized students for the benefit of privileged learners. Critical race pedagogy scholars suggest diversifying the curriculum to increase criticality and transform academic spaces into more inclusive environments (O'Neill & Miller, 2015). Educators must not only examine what they teach but also how they teach. Danowitz and Tuitt (2011) transformed their classroom by analyzing it as a space of liberation with the potential to impact individuals' lives, communities, and society as a whole. Building on the notion of "engaged pedagogy" by hooks (1994), these scholars embraced a reflexive and experiential approach aligned with feminist and critical pedagogies. Through autobiographical journaling, students can deeply examine their own racial and gender identity (Harbin et al., 2019). Harbin et al. (2019) encourage student reflectivity so that they can critically view their own social positions. Educators can also interrogate their own experiences of marginalization or privilege and share these reflections with students.

Many scholars who use critical race pedagogical approaches rely on critical race hypos (CRH) as a pedagogical tool to engage students in discussions about the link between their positionalities, personal experiences, and larger systems of oppression that mediate them (Pérez Huber & Solorzano, 2018). Through CRH, students are engaged in discussions which focus on the implications of microaggressions in social work classrooms and race, but also extend to other social locations and intersectional identities that are meaningful for racialized people. Focusing on microaggressions that trigger difficult dialogs and encounters in the classroom, Sue et al. (2009) suggest strategies to facilitate difficult discussions. It is important that educators validate feelings experienced by racialized students, legitimize a different racial reality, and exhibit good communication skills. However, when professors are unaware of racial dynamics, uncomfortable with race conversations, or ignore or dismiss race issues, the consequences could be quite devastating to racialized students.

CRH also helps students to make connections between everyday racism experienced by racialized people and larger systems of oppression. Interrogating race and whiteness in the classroom often poses challenges for educators (Harbin et al., 2019). Classroom space mirrors the color blind discourse in society that leads to resistance to and denial of race issues (Simpson et al., 2007). For example, Rothschild (2003), state that one of the biggest obstacles she encountered in teaching a predominantly white class about race was students' defensive responses and resistance to thinking of themselves as belonging to a racial group. Yet once this was accomplished, white students were more likely to be able to understand racial oppression. In predominantly white classrooms, she started a discussion about whiteness and dominant culture before addressing racism. According to Estrada and Matthews (2016), a course curriculum focused on race, racism, and other relevant topics can often trigger an emotional reaction among white students. Within anti-racist pedagogy, students unpack concepts such as cultural dominance, imperialism, and white racial privilege; this often results in white guilt or shame. However, white guilt can be a facilitative force in critical education at the postsecondary level.

Sensoy and DiAngelo (2014) explain how this pedagogy guides students in critical analysis of the presentation of mainstream knowledge as neutral, universal, and objective. It can also guide students in critical self-reflection of their own socialization into structured relations of oppression and privilege. Social justice pedagogy helps students develop analytical skills to challenge relations of power and oppression. Thus, critical social justice pedagogies develop strategies in classrooms that are responsive to omitted histories and positionalities. Critical race pedagogies emphasize the racialized identities and experiences of both educators and racialized students (Harbin et al., 2019; O'Neill & Miller, 2015).

Discussion and implications for social work

Teaching is inherently political; as such, educators should intentionally strive to develop a critical awareness of, and sensitivity to, the ways in which power and privilege unfold. It also means educators understanding their own and their students' positional power. Critical race scholarship has helped to expose how majoritarian structures have historically shaped and framed educational access and opportunities for historically marginalized and underrepresented populations. These marginalized groups are continuously overlooked and/or dismissed in higher education (Ledesma & Calderón, 2015). Pedagogical approaches guided by critical race theory pay specific attention to oppression, privilege, and relations of power while centering the voices of those who have been historically excluded.

In light of the proliferation of anti-Black racism, anti-Asian racism, anti-Indigenous racism, anti-Muslim racism and various forms of racism targeting racialized people, social workers have an ethical responsibility to challenge racial injustices. The social work profession and social work education remain mired in institutional systems of inequality. As social work is an engaged practice discipline that is positioned to examine daily intersectional injustices, utilizing liberatory and transformative critical race pedagogical approaches will "cast a new gaze on the persistent problems of racism" in society (Ladson-Billings & Tate, 1995, p. 60). In fact, social workers are uniquely positioned to address racial injustices because they work at the intersections of complex systems of service and care, and are called upon to ensure the accessibility of services for their clientele, many of whom are marginalized and racialized (Sulaimon et al., 2020). Critical race pedagogy is beneficial to new and experienced social workers to develop critical thinking skills through facilitating an understanding of how they engage with race and racism, and how this, in turn, impacts the ways they interact with racialized people.

Conclusion

Despite the lure of race-neutrality , colorblindness, and proclamations of a post racial society, substantial evidence reveals that the legacy of whiteness and white supremacy prevails. Racialized groups often experience systemic and institutional barriers that are further compounded by racism. Yet their experiences are often minimized or rendered invisible through various systemic processes of exclusion and erasure. Critical scholars have called for increased attention to the effects of systemic racism (Duhaney & El-Lahib, 2021; Mbakogu et al., 2021) Indeed, critical pedagogies that address issues of racism and power are needed to decenter the dominance of white privilege in the curriculum. CRT plays a distinct and critical role in examining issues of power and dominance. Integrating principles of critical race pedagogy are essential to naming inequities and devising strategies and practices to address them. Educators interested in understanding how they might integrate critical approaches in the classroom will find that critical race pedagogy informs social work curriculum, affirms the voices of racialized social work professors and racialized students, centers the deconstruction of dominant narratives that sustain racial oppression, and emphasizes reflexivity as a way to mitigate bias (Nakaoka & Ortiz, 2018). In particular, it is an important analytical tool to effectively articulate issues related to race and racism and provide opportunities for students to engage in courageous conversations that examine structural inequities. By forefronting critical race pedagogies, educators can portray the stories and experiences of racialized people as legitimate and valid. Beyond social work, "critical race pedagogy has the potential to unify existing critical explications of educational phenomena ... to provide more theoretical grounding and direction for educators who are concerned with issues of racial, ethnic, and gender inequality in the" (Lynn, 1999, p. 622), while creating more inclusive policies and practices.

Acknowledgments

The authors wish to thank Ebenezer Belayneh for contributing to the literature review and Ebony Morris for reviewing the manuscript.

Disclosure statement

No potential conflict of interest was reported by the author(s).

Funding

Funding was provided by the Taylor Institute for Teaching and Learning at the University of Calgary

ORCID

Liza Lorenzetti http://orcid.org/0000-0003-3791-4183

Prior publications

This paper has not been published online or in print and is not under consideration elsewhere.

References

Abrams, L. S., & Moio, J. A. (2009). Critical race theory and the cultural competence dilemma in social work education. *Journal of Social Work Education, 45*(2), 245–261. https://doi.org/10.5175/jswe.2009.200700109

Alemán, S. M., & Gaytán, S. (2017). "It doesn't speak to me": Understanding student of color resistance to critical race pedagogy. *International Journal of Qualitative Studies in Education, 30*(2), 128–146. https://doi.org/10.1080/09518398.2016.1242801

Alfred, T. (2009). *Wasáse: Indigenous pathways of action and freedom.* University of Toronto Press.

Allen, U. D. (2020). *COVID-19 among racialized communities: Unravelling the factors predictive of infection and adverse outcomes.* The Royal Society of Canada. https://rsc-src.ca/en/covid-19/impact-covid-19-in-racialized-communities/covid-19-among-racialized-communities-unravelling

Asher, N. (2009). Writing home/decolonizing text(s). *Discourse: Studies in the Cultural Politics of Education, 30*(1), 1–13. https://doi.org/10.1080/01596300802643033

Baines, D. (2011). *Doing anti-oppressive practice: Social justice social work.* Fernwood Publishing.

Banks, S., Cai, T., de Jonge, E., Shears, J., Shum, M., Sobočan, A. M., Strom, K., Truell, R., Úriz, M. J., & Weinberg, M. (2020). Practising ethically during COVID-19: Social work challenges and responses. *International Social Work, 63*(5), 569–583. https://doi.org/10.1177/0020872820949614

Barak, A. (2019). Critical questions on critical social work: Students' perspectives. *British Journal of Social Work, 49*(8), 2130–2147. https://doi.org/10.1093/bjsw/bcz026

Bauer, M. (2000). An essay review: Implementing a liberatory feminist pedagogy: Bell hooks' strategies for transforming the classroom. *MELUS, 25*(3/4), 265–274. https://doi.org/10.2307/468246

Blakeney, A. M. (2005). Antiracist pedagogy: Definition, theory, and professional development. *Journal of Curriculum and Pedagogy, 2*(1), 119–132. https://doi.org/10.1080/15505170.2005.10411532

Brown, C. G. (2012). Anti-oppression through a postmodern lens: Dismantling the master's conceptual tools in discursive social work practice. *Critical Social Work, 13*(1), 34–65. https://doi.org/10.22329/csw.v13i1.5848

Butler, A., Elliott, T., & Stopard, N. (2003). Living up to the standards we set: A critical account of the development of anti-racist standards. *Social Work Education, 22*(3), 271–282. https://doi.org/10.1080/0261547032000083469

Campbell, E. (2014). Using critical race theory to measure "racial competency" among social workers. *Journal of Sociology and Social Work, 2*(2), 73–86. https://doi.org/10.15640/jssw.v2n2a5

Clarke, J. (2011). The challenges of child welfare involvement for Afro-Caribbean families in Toronto. *Children and Youth Services Review, 33*(2), 274–283. https://doi.org/10.1016/j.childyouth.2010.09.010

Cooper, A. (2006). *The hanging of Angelique: The untold story of Canadian slavery and the burning of Old Montreal.* Harper Perennial.

Coxshall, W. (2020). Applying critical race theory in social work education in Britain: Pedagogical reflections. *Social Work Education, 39*(5), 636–649. https://doi.org/10.1080/02615479.2020.1716967

Crenshaw, K., Gotanda, N., Peller, G., & Thomas, K. (Eds.). (1995). *Critical race theory: The key writings that formed the movement.* The New Press.

Danowitz, M. A., & Tuitt, F. (2011). Enacting inclusivity through engaged pedagogy: A higher education perspective. *Equity & Excellence in Education, 44*(1), 40–56. https://doi.org/10.1080/10665684.2011.539474

Dei, G. J. S. (2000). Towards an anti-racism discursive framework. In M. Aguiar, M. Calliste, & G. J. S. Dei (Eds.), *Power, knowledge & anti-racism education: A critical reader* (pp. 23–40). Fernwood Publishing.

Dominelli, L. (2002). *Anti-oppressive social work theory and practice.* Palgrave Macmillan.

Duhaney P and El-Lahib Y. (2021). The Politics of Resistance From Within. *ASW, 21*(2/3), 421–437. https://doi.org/10.18060/24471

Elias, M. J., Zins, J. E., Weissberg, R. P., Frey, K. S., Greenberg, M. T., Haynes, N. M., Kessler, R., Schwab-Stone, M. E., & Shriver, T. P. (1997). Promoting social and emotional learning: Guidelines for educators. *Association for Supervision and Curriculum Development.* https://earlylearningfocus.org/wp-content/uploads/2019/12/promoting-social-and-emotional-learning-1.pdf

Esposito, L., & Romano, V. (2016). Benevolent racism and the co-optation of the Black Lives Matter movement. *The Western Journal of Black Studies, 40*(3), 161–173. https://www.proquest.com/docview/2049976066

Estrada, F., & Matthews, G. (2016). Perceived culpability in critical multicultural education: Understanding and responding to race informed guilt and shame to further learning outcomes among White American college students. *International Journal of Teaching and Learning in Higher Education, 28*(3), 314–325. https://files.eric.ed.gov/fulltext/EJ1125096.pdf

Fortier, C., & Wong, E. H. S. (2018). The settler colonialism of social work and the social work of settler colonialism. *Settler Colonial Studies, 9*(4), 437–456. https://doi.org/10.1080/2201473X.2018.1519962

Freire, P. (1970). *Pedagogy of the oppressed.* Seabury Press.

Goulet, S. (2018). From racism to reconciliation: Indigenous peoples in Canada. In D. Este, L. Lorenzetti, & C. Sato (Eds.), *Racism and anti-racism in Canada.* Fernwood Publishing. 74–101.

Hager, M. (2020, April 22). *Vancouver sees surge in hate crimes against East Asian people*. The Globe and Mail. https://www.theglobeandmail.com/canada/british-columbia/article-vancouver-sees-surge-in-hate-crimes-against-east-asian-people/

Harbin, M. B., Thurber, A., & Bandy, J. (2019). Teaching race, racism, and racial justice: Pedagogical principles and classroom strategies for course instructors. *Race and Pedagogy Journal: Teaching and Learning for Justice, 4*(1), 1–37. https://archives.pdx.edu/ds/psu/29958

hooks, B. (1994). *Teaching to transgress: Education as the practice of freedom*. Routledge.

Hyman, I., & Vissandjee, B. (2020). COVID-19, intersectionality and concerns about violence at home among immigrant men and women. *Canadian Diversity, 17*(3), 34–40. https://scholar.google.com/scholar_lookup?title=Covid-19%2C%20Intersectionality%20and%20Concerns%20about%20Violence%20at%20Home%20among%20Immigrant%20Men%20and%20Women&journal=Canadian%20Diversity&volume=17&issue=3&pages=34-40&publication_year=2020&author=Hyman%2CI&author=Vissandjee%2CB

Jedwab, J. (2020). *Canadian opinion on the coronavirus – No. 14: Economic vulnerability score for selected visible minorities and the effects of COVID-19*. Association for Canadian Studies. https://acs-aec.ca/wp-content/uploads/2020/04/ACS-Covid-and-Economic-Vulnerability-of-Visible-Minorities-April-2020.pdf

Josephine, E., & Hyman, I. (2021). Unpacking the health and social consequences of COVID-19 through a race, migration and gender lens. *Canadian Journal of Public Health, 112*(1), 8–11. https://doi.org/10.17269/s41997-020-00456-6

Keating, F. (2000). Anti-racist perspectives: What are the gains for social work? *Social Work Education, 19*(1), 77–87. https://doi.org/10.1080/026154700114676

Khare, N., Shroff, F., Nkennor, B., & Mukhopadhyay, B. (2020). Reimagining safety in a pandemic: The imperative to dismantle structural oppression in Canada. *Canadian Medical Association Journal, 192*(41), E1218–E1220. https://doi.org/10.1503/cmaj.201573

Kishimoto, K. (2018). Anti-racist pedagogy: From faculty's self-reflection to organizing within and beyond the classroom. *Race, Ethnicity and Education, 21*(4), 540–554. https://doi.org/10.1080/13613324.2016.1248824

Ladson-Billings, G., & Tate, W. F., IV. (1995). Towards a critical race theory of education. *Teachers College Record, 97*(1), 47–68. https://doi.org/10.1177/016146819509700104

Laurencin, C. T., & Walker, J. M. (2020). A pandemic on a pandemic: Racism and COVID-19 in Blacks. *Cell Systems, 11*(1), 9–10. https://doi.org/10.1016/j.cels.2020.07.002

Ledesma, M. C., & Calderón, D. (2015). Critical race theory in education. *Qualitative Inquiry, 21*(3), 206–222. https://doi.org/10.1177/1077800414557825

Lindeman, T. (2021, May 28). *Canada: Remains of 215 children found at Indigenous residential school site*. The Guardian. https://www.theguardian.com/world/2021/may/28/canada-remains-indigenous-children-mass-graves

Louis-Jean, J., Cenat, K., Njoku, C. V., Angelo, J., & Sanon, D. (2020). Coronavirus (COVID-19) and racial disparities: A perspective analysis. *Journal of Racial and Ethnic Health Disparities, 7*(6), 1039–1045. https://doi.org/10.1007/s40615-020-00879-4

Lupton, A., & Dubinski, K. (2021, June 8). *What we know about the Muslim family in the fatal London, Ont., truck attack*. CBC News. https://www.cbc.ca/news/canada/london/london-muslim-family-attack-what-we-know-1.6057745

Lynn, M. (1999). Toward a critical race pedagogy: A research note. *Urban Education, 33*(5), 606–626. https://doi.org/10.1177/0042085999335004

Lynn, M. (2004). Inserting the "race" into critical pedagogy: An analysis of "race based Epistemologies. *Educational Philosophy and Theory, 36*(2), 153–165. https://doi.org/10.1111/j.1469-5812.2004.00058.x

Mallapragada, M. (2021). Asian Americans as racial contagion. *Cultural Studies, 35*(2–3), 279–290. https://doi.org/10.1080/09502386.2021.1905678

Maynard, R. (2017). *Policing Black lives: State violence in Canada from slavery to the present*. Fernwood Press.

Mbakogu I, Duhaney P, Ferrer I and Lee E Ou. Confronting whiteness in social work education through racialized student activism. Canadian Social Work Review, 38(2), 113-140 10.7202/1086122ar

McCoy, D. L., & Rodricks, D. J. (2015). Critical race theory in higher education: 20 years of theoretical and research innovations. *ASHE Higher Education Report, 41*(3), 1–117. https://doi.org/10.1002/aehe.20021

Mullaly, R. (1997). *Structural social work: Ideology theory and practice* (2nd ed.). Oxford University Press.

Nakaoka, S., & Ortiz, L. (2018). Examining racial microaggressions as a tool for transforming social work education: The case for critical race pedagogy. *Journal of Ethnic & Cultural Diversity in Social Work, 27*(1), 72–85. https://doi.org/10.1080/15313204.2017.1417947

Noack, A. M., & Vosko, L. F. (2011). *Precarious jobs in Ontario: Mapping dimensions of labour market insecurity by workers' social location and context*. Law Commission of Ontario. https://www.lco-cdo.org/wp-content/uploads/2012/01/vulnerable-workers-call-for-papers-noack-vosko.pdf

O'Neill, P., & Miller, J. (2015). Hand and glove: How the curriculum promotes an antiracism commitment in a school for social work. *Smith College Studies in Social Work, 85*(2), 159–175. https://doi.org/10.1080/00377317.2015.1021222

Ontario Human Rights Commission. (2018). *A collective impact: Interim report on the inquiry into racial profiling and racial discrimination of Black persons by the Toronto Police Service.* http://www3.ohrc.on.ca/sites/default/files/TPS%20Inquiry_Interim%20Report%20EN%20INAL%20DESIGNED%20for%20remed_3_0.pdf

Ortiz, L., & Jani, J. (2010). Critical race theory: A transformational model for teaching diversity. *Journal of Social Work Education, 46*(2), 175–193. https://doi.org/10.5175/JSWE.2010.200900070

Parker, L., & Stovall, D. O. (2004). Actions following words: Critical race theory connects to critical pedagogy. *Educational Philosophy and Theory, 36*(2), 167–182. https://doi.org/10.1111/j.1469-5812.2004.00059.x

Pérez Huber, L., & Solorzano, D. G. (2018). Teaching racial microaggressions: Implications of critical race hypos for social work praxis. *Journal of Ethnic & Cultural Diversity in Social Work, 27*(1), 54–71. https://doi.org/10.1080/15313204.2017.1417944

Rothschild, T. (2003). "Talking race" in the college classroom: The role of social structures and social factors in race pedagogy. *Journal of Multicultural Counseling and Development, 31*(1), 31–38. https://doi.org/10.1002/j.2161-1912.2003.tb00528.x

Sagris, J. (2008). Liberatory education for autonomy. *The International Journal of Inclusive Democracy, 4*(3), 1–4. https://www.inclusivedemocracy.org/journal/vol4/vol4.htm

Sensoy, Ö., & DiAngelo, R. (2014). Respect differences? Challenging the common guidelines in social justice education. *Democracy and Education, 22*(2), 1–10. https://democracyeducationjournal.org/home/vol22/iss2/1

Shadmi, E., Chen, Y., Dourado, I., Faran-Perach, I., Furler, J., Hangoma, P., Hanvoravongchai, P., Obando, C., Petrosyan, V., Rao, K. D., Ruano, A. L., Shi, L., de Souza, E. L., Spitzer-Shohat, S., Sturgiss, E., Suphanchaimat, R., Villar, M. U., & Willems, S. (2020). Health equity and COVID-19: Global perspectives. *International Journal Equity Health, 19*(104), 1–16 . https://doi.org/10.1186/s12939-020-01218-z

Simpson, J. S., Causey, A., & Williams, L. (2007). "I would want you to understand it": Students' perspectives on addressing race in the classroom. *Journal of Intercultural Communication Research, 36*(1), 33–50. https://doi.org/10.1080/17475750701265274

Sue, D. W., Lin, A. I., Torino, G. C., Capodilupo, C. M., & Rivera, D. P. (2009). Racial microaggressions and difficult dialogues on race in the classroom. *Cultural Diversity & Ethnic Minority Psychology, 15*(2), 183–190. https://doi.org/10.1037/a0014191

Sulaimon, G., Mullings, D. V., Adjei, P. B., & Karki, K. K. (2020). Racial erasure: The silence of social work on police racial profiling in Canada. *Journal of Human Rights and Social Work, 5*(4), 224–235. https://doi.org/10.1007/s41134-020-00136-y

Suoranta, J., & Moisio, O. P. (2006). Critical pedagogy as collective social expertise in higher education. *International Journal of Progressive Education, 2*(3), 47–64. http://www.inased.org/IJPEv2n3.pdf

Thomas Bernard, W. (2020). When two pandemics collide: Racism, COVID-19 and the Association of Black Social Workers emergency response. *Canadian Social Work Review, 37*(2), 175–183. https://doi.org/10.7202/1075119ar

Treviño, A. J., Harris, M. A., & Wallace, D. (2008). What's so critical about critical race theory? *Contemporary Justice Review, 11*(1), 7–10. https://doi.org/10.1080/10282580701850330

Truth and Reconciliation Commission of Canada. (2015). *Honouring the truth, reconciling for the future: Summary of the final report of the truth and reconciliation commission of Canada.* https://ehprnh2mwo3.exactdn.com/wp-content/uploads/2021/01/Executive_Summary_English_Web.pdf

United Nations. (2016). *UN expert panel warns of systemic anti-Black racism in Canada's criminal justice system.* https://news.un.org/en/story/2016/10/543482-un-expert-panel-warns-systemic-anti-black-racism-canadas-criminal-justice

Williams, C. (1999). Connecting anti-racist and anti-oppressive theory and practice: Retrenchment or reappraisal? *British Journal of Social Work, 29*(2), 211–230. https://doi.org/10.1093/oxfordjournals.bjsw.a011443

Williams, C. (2019). Critical social work in the new urban age. In S. Webb (Ed.), *The Routledge handbook of critical social work* (pp. 267–277). Routledge. https://doi.org/10.4324/9781351264402

Coloniality of power, critical realism and critical consciousness: the three "C" framework

Lisa Werkmeister Rozas ⓘ

ABSTRACT

Typical pedagogical practices center Whiteness, particularly when teaching about racism and racial justice. This article offers a framework that de-centers the White frame using: coloniality of power, critical realism, critical consciousness. The coloniality of power analyzes the order of social relations and embedded hegemonic structures in the US. Critical realism posits a multi-layered construct of reality encompassing subjugated experiences, illustrating how dominant groups can share these experiences. Critical consciousness explains the need for all individuals to identify mechanisms of oppression that maintain, perpetuate, and sustain the exclusion and sub-jugation of BIPOC. Together, these three concepts create a foundation highlighting accountability and agency.

The term racial justice has found its way into social, academic, and political settings and as a result has become a placeholder for all things having to do with racism, racial discrimination, racial prejudice, racial health and racial economic inequities. With the ongoing appalling murders of black men and women by police, most recently of George Floyd and Breonna Taylor, the public's attention is fixated on race. Due to the legacy of slavery, the genocide of the indigenous cultures, and the continuous benefit White individuals proffer as a result, people born and raised in the US are socialized to see everything through the lens of race. However, depending on one's own lived experiences, how race is actually lived, experienced and perceived, varies as does its consequences. Building an awareness of how these issues and experiences intersect is the first step in developing a critical response to dismantling racial injustice. In order to participate in the important work of achieving racial justice, we must provide people with information about levels, systems, and structures of racism and opportunities to process this information in an experiential way. This paper offers a pedagogical framework consisting of three C's: Coloniality of Power, Critical Realism and Critical Consciousness, each, when utilized together, serve to facilitate a path for much needed advocacy and activism for racial justice.

Few would argue that White supremacy has relegated race to be the dominant paradigm in the United States. However, this ideology alone is not the sole cause of the unabashed injustice endured by Black Indigenous People of Color (BI-POC). The COVID-19 pandemic has illuminated this injustice all too well. Studies have shown that Black, Asian, and Latinx individuals were at greater risk of contracting the virus and had worse clinical outcomes than Whites (Pan et al., 2020). What cannot be ignored and must be seen together in the ongoing dehumanization of BI-POC, are four arcane and fervent forces that support, embolden, strengthen and perpetuate the racism and racial injustice that continues to cripple the United States. Those forces and their seemingly indissoluble network are outlined in Quijano's Coloniality of power (2007): White supremacy, patriarchy, capitalism and Christianity. Only by attacking these four axes can the systems that bolster and harbor racism be dismantled. Crucial to engaging in anti-racist praxis is understanding the matrix of coloniality, including how it can be illuminated, is experienced, and can be abolished. This paper will explain the four power sources outlined in the coloniality of power, the utility of critical realism to understand differential experiences within this power structure and how, together, they can be used to develop students' critical consciousness. This explanatory framework outlined as the three 'C's help to inform and propel the dismantling of the vicious matrix of coloniality and anti-racist praxis of all kinds.

Coloniality of power

Although not extinct, colonialism as it was once known has declined and is admonished by most nation states. Most modern examples comprise territories owned by the US government that do not share sovereignty (e.g., Puerto Rico, Guam, American Samoa) and Indigenous tribes whose land has been stolen, governance structures decimated, resources extracted and overall ability to "self-rule . . . ultimately subject to limits set by the colonial power" (Bacon & Norton, 2019, p. 304). What also persists is a residue that although invisible continues to be a potent force in the socio-political and economic oppression of targeted racial groups in contemporary society. Quijano's, 2007) concept of the coloniality of power is one that addresses these "colonial situations" that continue to exist in the absence of a true colonial administration. He contends that the matrix of coloniality, evidenced through the codification of race and subsequent racial social classification, the distribution and valuation of work based on racial identities, the racialization of certain cultural identities, the hoarding of resources and capital, the regulation of social relations, has as its core the power and privilege of Whiteness (218). Basically, race and racism become the organizing principles of how hierarchies are structured in the modern world (Grosfoguel, 2011). Such hierarchical structures have been laid bare during the COVID-19 pandemic, particularly when examining which segments of the workforce had the privilege of sheltering in place and decreasing their risk of contracting the virus. Low-income Black, Latinx, Indigenous, and other minoritized ethnic groups in large part, did not have the privilege to work from home (Xu & Li, 2020).

According to Grosfoguel, patriarchal valuation of gender is also affected by this as some women who would be classified as "White" in some circumstances are granted superiority over men who would be classified as "non-White." Race becomes the lens through which all people are ascribed value. Grosfoguel adds to Quijano's matrix by contending that "race, gender, sexuality, spirituality, and epistemology are not additive elements to the economic and political structures of the capitalist world-system, but an integral, entangled and constitutive part of the broad webbed "package" called the European modern/colonial capitalist/patriarchal world-system" (Grosfoguel, 2011, p. 11).

Historically, Christianity has supported and has been supported by White supremacy. Hill Fletcher explains, "The systems and structures have been intimately joined with Christian supremacy, such that, undoing White supremacy will also require relinquishing the ideologies and theologies of Christian supremacy (2017). The belief that Christians were anointed to carry out a destiny that was pre-ordained by God sanctified their role as liberators to all who were not yet part of humanity that Christ imagined. This is particularly evident in Europe during Medieval times when Christian crusades were mounted to stop the expansion of Islam, killing thousands. Similarly, not only did Christian missionaries all but eradicate the cultural and spiritual practices of myriad indigenous peoples they have been implicated in supporting the genocide of various indigenous tribes in North and South America. Both campaigns also included the acquisition of land (The Pluralism Project, Harvard University). Their desire to conquer and deliver the White racial frame (Feagin, 2020) ensconced in Christianity would become one of the foundations and cross beams supporting the devaluation and exploitation of all BI-POC. Less attention is being paid to the role Christianity has in subsidizing racism. However, it has become more salient after four years of a Trump presidency.

What has also become more visible is the upsurge in Christian-Nationalists whose values align with nativism, White supremacy, patriarchy, and heteronormativity (Whitehead & Perry, 2020). Many USAmericans who uphold Christianity as a pathway to patriotism utilize examples of United States laws and formative documents having biblical origins and the people rewarded by God (Gorski, 2019). One must only look at the behavior of certain White supremacist groups to recognize its power. Before participating in the siege on the US Capital on January 6, 2021, members of the Proud Boys bowed their heads and said a prayer. This extremist group has been described as maintaining values that support anti-Black racism, misogyny, xenophobia and overall hatred toward people of color. It would be important to acknowledge that the majority of Christians in the United States do not consciously espouse nationalist values yet what must be conceded is that for some, Christianity is a mantle used to exculpate bigotry and prejudice.

Husain (2017), asserts that race and religion are two strong forces that have shaped USAmerican society, particularly when it comes to allotment of resources, opportunity, domination, and subjuga-tion. However, it is important to acknowledge that there is a distinction between White and Black Christianity. Historically, Black churches played a pivotal role in the lives of many Blacks who were enslaved. They were one of the few places where Black voices, identification, and culture could be promulgated. They provided a place for emotional and social support as well as affording them a space to resist and mobilize against White supremacy (Allen, 2019). However, Christianity has also been known to support inequality by encouraging its followers to focus on attaining justice and salvation in the next world subjugating Black collective agency (Morris, 2001).

By situating individuals within the context of these power structures we are able to more fully analyze how they experience society and how society experiences them. All individuals are classed, sexed, gendered, and racialized; each identity carries with it its own varying levels of power and privilege, which makes up our social location (Garran & Werkmeister Rozas, 2013). Grosfoguel makes the point that individuals whose social location comprise social categories that primarily carry less power and privilege may possess a subjugated view of their own identities. Bell and Griffin (2007) identify the espousal of the oppressor's world views and beliefs by the oppressed as internalized oppression.

True knowledge has been established by those who hold power within the matrix as well as how that knowledge is imparted. In this regard the validity and scope of subjugated peoples' knowledge and experience has long been deemed inferior. *How* we learn and *what* we learn has been dictated by an epistemology steeped in the matrix of coloniality. Though when included in the *what* is to be studied, the experiences of BIPOC are conveyed through a lens of coloniality, locating them outside the worthy standard.

Most important to keep in mind is, as Grosfoguel states, "The intersectional entanglement between class, sexual, gender or national/colonial oppressions that exist in the zone of non-being are ... qualitatively distinct from the ways these oppressions are lived and articulated in the zone of being" (2016, p. 12). Therefore, fundamental to an intersectional analysis of social categories within the United States is an a priori acknowledgment of the existence and distinct difference that these two zones create for their inhabitants. Individuals who experience oppression because of their class, gender or sexuality but have White privilege inhabit the zone of being and are mutually recognized for their humanity within the coloniality of power. However, their targeted or marginalized social identity places them in opposition to the dominant gendered, classed, and/or sexual hegemonic groups and for this they become the "Other Being." When a racially oppressed individual inhabits one of these targeted or marginalized identities that conflict with the hegemony they become the "Non-Being Other" with no mutual recognition of humanity and no race privilege to mitigate their position (Grosfoguel, 2016, p. 14).

Critical realism

The question of whose reality is valid and how one goes about the process of determining experience from perception is relevant when discussing racist, discriminatory, and/or prejudicial events. Social work, as a profession, has been complicit in supporting White dominance through centering the White experience and privileging individualism. The marginalization of BI-POC voices and invalidation of their reality by leaders of the profession, has been identified and slowly acknowledged (Wright et al., 2021). What has yet to be addressed is how the embedded ideology of White Supremacy, which privileges individualism, continues to obfuscate the causal forces that shape, maintain, and perpetuate racial injustice. This omission itself has causal powers that generate doubt and diffidence among BI-POC when they endeavor to make sense of their reality. Many BI-POC go through a series of questioning when confronted with microaggressions. In attempting to ascertain how to interpret a comment or action that is covertly racist, BI-POC often participate in a kind of "mental math" (Werkmeister Rozas & Miller, 2009, p. 35); their minds constantly adding, subtracting, and re-adding what they know of themselves and the world they live in to be certain of what really occurred. Their process is one that includes constant questioning of the basis for, and how to interpret a microaggressive act. Individuals with dominant social identities not only are spared this exhausting routine, because they experience the world as the world experiences them, they are not even aware of the process BIPOC undertake.

Critical realism was developed by Roy Bhaskar as a philosophical approach to the natural sciences, to counter both positivism and postmodernism (Archer et al., 1998). Its principles have since been applied to a number of different disciplines including ethics and social science. It has also been identified as a meta-theory that can be used in the "Indigenous decolonizing processes" (Hockey, 2010, p. 368). Critical realism posits that an objective reality does exist, and it is composed of three levels: "the *empirical* level consisting of experienced events; the *actual* level, comprising all events whether experienced or not; and, lastly the *causal* level, embracing the 'mechanisms' which generate events" (Houston, 2001, p. 850). Understanding that the causal level is still real, though not always perceptible, is crucial, as it affects the events occurring on all levels. Archer et al. (1998) provides the concept of magnetism to illustrate the causal level, and its ability to affect the empirical- a mechanism that, though unseen, has real influence. Critical realism also describes the world as a set of systems whose influence on one another is variable, significant (or insignificant), and unpredictable. Although tendencies may exist, these systems are all operating simultaneously (Houston, 2001) and without a set pattern. Human systems are described as complex due to the countless social and psychological factors that at any given time and every given moment affect individuals and groups.

Using this lens, oppression can be seen as a system occurring at the causal level as well as the empirical; it is a force that has significant influence on all facets of the human experience. Hegemonic systems such as White supremacy, patriarchy, Christian-centrism, are *unseen mechanisms* that construct the *empirical* and *actual* levels of experience such as discrimination, racism, sexism, Islamophobia and anti-Semitism. Often, we presume that unseen oppressive forces influence in one direction only; the oppressor impacting the oppressed. Houston (2001) makes clear that, when applied as a social theory, critical realism does not presume individuals to be merely victims of these unseen mechanisms, without recourse. Rather, just as these forces impact us, we can impact them.

The purpose of exposing and understanding oppression is to change it, to work toward the "emancipatory transformation of those structures" (Houston, citing O'Neill). A core activity of critical consciousness development, therefore, is to progress from discursive awakening to social action (Freire, 1973). Critical realism can inform this process. Recast in critical realist terms, conscienticization can be understood as developing an analytic lens with which we assess our empirical experience in the context of actual and causal levels of the coloniality of power to advance the transformative capacity of self-emancipation (Hockey, 2010).

Critical consciousness

Critical consciousness is the ability to recognize political, social and economic forms of oppression and critically engage in dismantling them. Systematically understanding the role of oppression in shaping one's reality is therefore essential to the social transformation process. As described by Freire (2018) critical consciousness involves an awareness of the following forces: power, critical discourse, human agency. This enables the students to examine not only power dynamics that exist on the individual level but also the transformation of institutional policies and practices on a structural level. Recognizing the mechanisms that operate on a structural level helps students to understand the interconnection between oppressive ideologies, social systems, and interpersonal interactions. In other words, only looking at racism, sexism and other forms of oppression as interpersonal transactions or as individual experiences misses the influence of durable social, political, and economic structures undergirding social hierarchies. Simultaneously, only considering oppression as the result of abstract clandestine systems can obscure the ways in which we create and recreate forms of oppression through our actions and inactions, as well as diminish the role individual human agency has in confronting change. Being aware of the societal structures that perpetuate and maintain racism is one aspect of critical consciousness. Houston (2001) writes that "One of the central challenges in the social and psychological sciences at the present time is how to promote a theory of the human agency whilst taking into account the impact of social structure" (p. 849). The "impact of social structure" refers to those causal forces that occur in the unseen realm yet have an empirical footprint. Consider the research that identifies BIPOC individuals are at greater risk of COVID-19 infection, hospitalization and death (Pan et al., 2020). Having this empirical data suggests that this disparity exists, yet many of the causal forces, such as lack of social distancing conditions for low-wage workers, reliance on public transportation, and congested living areas remain in the unseen realm. Agency, then, is the individual's ability to become critically aware of these forces and factors, and work to reconcile and mitigate their influence. Exposing the causal factors that put BIPOC individuals at greater risk for COVID-19 and developing solutions to interrupt these forces is the goal. Drawing attention to, and educating people about, the systemic realities and functions of oppression, through the development of a critical consciousness (Freire, 2018), is one of the main goals of racial justice. Critical consciousness is developed through identification and acknowledgment of the current social and historical context that illustrate the magnitude and depth of oppressive systems influencing people's lived experiences (Freire, 1970). The systemic focus is especially important for people who may have less of an understanding of privilege and oppression. Understanding critical consciousness development is important for understanding and engaging in the fight for racial justice.

Critical consciousness development and critical realist conceptualization of oppression

Emphasizing critical consciousness development in this way may help social workers, and all who are pursuing a praxis of critical consciousness, transcend the common focus on the individual level of oppression, while grounding an abstract conceptualization of systemic oppression in the empirical world of institutions and human agents. Critical consciousness development through a critical realist lens allows us to make connections between our empirical experiences, the empirical experiences of others, and actual and causal levels of systemic oppression. Freire (1973) believed strongly in the role critical consciousness plays in our ability to experience what is real, "the more accurately men [sic] grasp true causality, the more critical their understanding of reality will be" (44). It also offers a framework for transforming these systems.

Multi-level oppression and critical realism

To apply critical realism to the critical consciousness development process it is necessary to first describe oppression in critical realist terms, i.e. as comprising three levels: the causal, actual and empirical. At the causal level power structures serve as unseen forces which shape social life and co-opt difference to create social hierarchy. Such forces include those within the matrix of coloniality: White supremacy, patriarchy, capitalism and Christian-centrism. In the United States, these interrelated social suprastructures pervade all aspects of social, cultural, political and economic life. At the actual level these social forces are mechanized through institutions, interactions, self-perceptions and cultural norms. Privilege, microaggressions, cultural appropriation, residential segregation, and social marginalization are all examples of actual manifestations of the coloniality of power. These physical, socio-political, economic and cultural structures exist whether or not an individual actor consciously experiences them. We can recognize their existence through our empirical experiences, either personally or in aggregate. Personally, we have empirical experiences informed by oppression. Oppression is manifested when, for example, one lives in a neighborhood that has been decimated by economic and political divestment. It is present when one strives to attain a standard of beauty that sexually subjugates women and norms beauty to whiteness. It is also evident when accessing the benefits of privilege by, for example, engaging in self-deception associated with the myth of meritocracy, or believing that one's personal values are ratified by God. There are countless empirical examples of oppression that impact our everyday lives.

Empirical manifestations of oppression can also be visible at the aggregate level. The aggregate empirical level illustrates patterns across individual experience. For example, racial disparities in health, from the overrepresentation of BIPOC who have had and died of COVID-19 (Bibbins-Domingo, 2020) to the underrepresentation of BIPOC who are receiving the vaccine, (Schoenfeld Walker et al., 2021), reveal the collective effects of oppression on targeted groups, as do gender and race-based wealth and wage gaps. Mass incarceration of men of color provides empirical evidence of the aggregate consequences of systemic racism within the policies and practices of law enforcement and the criminal justice system (Alexander, 2010). The various levels of oppression when seen through a critical realist terms are seen as overlapping rather than nested because of the interconnected effects each level has on the other.

Tracing one specific form of oppression through the various levels explicated in critical realism may help articulate the framework. The COVID-19 pandemic has uncovered how much the United States relies on the workers that grow, harvest, and pack food. It has also shined a spotlight on the thousands of farm and factory workers that currently and for generations have lived and worked in sub-standard and unsafe conditions (Coalition of Immokalee workers: https://ciw-online.org/blog/2021/02/wendys-supply-chain-protections-covid/). At the empirical level, an exploited worker directly experiences oppression. She may be one of the 80,000 meat-packing, food processing and agricultural workers who have tested positive for COVID-19 or one of the over 1,000 who were among the excess deaths due to COVID-19 (Chen et al., 2021). Substandard housing conditions (Arcury et al., 2012; Quandt et al., 2013), exposure to toxic chemicals (Arcury et al., 2001; Robinson et al., 2011), limited bargaining power (Telega & Maloney, 2010), sexual harassment (Murphy et al., 2015; Waugh, 2010) and lack of job security (McLaughlin & Hennebry, 2013) are also common conditions in which workers find themselves. If she has access to aggregated information regarding labor abuses in the agriculture sector, she may see that the issues she has encountered do not just impact one or two farms. These issues are present at many farms in many locations (Farmworker Justice, 2017), particularly those that are large produce suppliers.

The actual level consists of the cultural norms, institutions and practices which solidify power in the hands of agribusiness leaders and away from the worker. This includes laws and policies that perforate the labor rights of farmworkers (Ramos, 2018) and racialized narratives that permit the marginalization of Latinx workers (Nelson, 2008). Capitalism also works at the actual level, creating fundamental structures which commodify labor and demand the extraction of surplus cheap labor of many for the profit of few.

Finally, the unseen mechanisms at work behind all of this are White supremacy, which denigrates nonwhite people and rationalizes their mistreatment; patriarchy, which undermines the credibility of the concerns of women (particularly women of color); Christian-nationalism which contends that Christians are the only true patriots and their mission is to rescue the United States from the godless (Whitehead & Perry, 2020). The overarching matrix of coloniality (Quijano, 2007), recasts indigenous land as a natural resource that was gifted to Evangelical Christians by God, to be manipulated for profit, augmenting the justification for exploitation. This gift should be taken back by those who deserve it (Gorski, 2019).

Critical realism is a scaffolding upon which to build an understanding of one's empirical experience in the context of systemic oppression. This can be accomplished by directing time and attention to tracing a personal empirical experience through the aggregate empirical, actual and causal levels which shape it. This practice of critical consciousness development unearths how oppression may shape one's experience in ways previously obscured. In the farmworker example above, going "up the ladder" of the different levels of experience would help the farmworker recognize -to the extent that she hasn't already- 1) the harm exploitation is currently doing to her (some of these things would be empirically obvious, such as working without proper personal protection equipment, others may not be so clear, such as exposure to chemicals, or subtle forms of wage theft); 2) patterns across other farms which reveal the systemic nature of exploitation and the collective interest of farmworkers across sites; 3) the institutions, policies and practices at the actual level and how they shape her life; and 4) the unseen forces of power which undergird the entire system. Connecting empirical experience to systemic forces of oppression can be a powerful way for targets of oppression to situate individual hardship within a collective struggle. Indeed, social movements have utilized this process to promote collective mobilization across a range of issues (see for example, Cornish, 2006).

Direct targets of oppression are not the only ones who can and should develop a critical consciousness. All actors in the system maintain and advance the oppression of the farmworker and therefore all have a role in dismantling it. Freire (1973) purports that all people are responsible for developing a greater awareness of their socio-political-cultural realities so they can engage in the analysis of the social forces that lead to transformation. The farm manager, farm owner, agribusiness lobbyist, and politician all play a role. However, it is the consumer who may best illustrate the utility of critical realism to assist in the critical consciousness development process for individuals who are, intentionally or unintentionally, in the "oppressor" group and hope to move toward ally-ship through consciousness development.

When I purchase poultry that was produced by way of exploitative labor practices, for example, I have an empirical experience that is related to the factory worker's oppression. I purchase an item that, due to exploitation, is less expensive than it might otherwise have been. However, the connection between my purchase and the experience of the factory worker is obscured. It is here that the concept of the *actual level* helps facilitate critical consciousness development in those with privilege. The actual level consists of realities that shape your experience *whether or not you recognize them or experience them yourself.* Understanding this level reveals a dimension of reality that had been previously shielded from sight through the blinders of power and privilege. Critical consciousness leads me to recognize that my seemingly innocuous act of purchasing a discounted item unwittingly implicates me in the system that oppresses the factory worker. I cannot act on what I did not know. However, once the connection to oppression has been brought into my consciousness (often by an external stimulus such as a public awareness campaign or boycott) my empirical experience is altered. I view the item not just for its immediate physical properties, but for what I now know it to be- connected, literally, physically, to the hands of an oppressed worker. "Oppression" is no longer a distant and abstract concept, but a reality tied to my own, of which I have been an active part. As I learn more about the oppression of this worker, I may begin to understand how it connects to larger systems of patriarchy, White supremacy, capitalism, the power of coloniality and, perhaps, how they obfuscate my other empirical experiences of privilege and oppression. This may reveal additional blind spots that obscure related actual level systems of oppression.

Patomäki and Wight (2000) discuss the limitations of relying on one's own perceptions to define reality for the self.

[A]ccording to critical realism the world is composed not only of events, states of affairs, experiences, impressions, and discourses, but also of underlying structures, powers, and tendencies that exist, whether or not detected or known through experience and/or discourse ... For both the underlying reality that makes experience possible and the course of events that is not experienced/spoken are reduced to what can be experienced or become an object of discourse. (p. 223)

Critical consciousness development is a way to bring evidence of oppression at the causal level into one's conscious experience thereby changing one's relationship with the empirical world.

Once my empirical experience is thus altered, I am accountable for my role in the system. I am confronted with a choice to act or maintain the status quo. Examining the actual level mechanisms that facilitate farmworker oppression provides insight into how to act. Practices, policies, and cultural norms are potential targets of intervention. For example, I can revise my purchasing decisions (for instance, purchasing products with documented fair labor practices), engage in an organized movement to support pro-farmworker policies, or participate in the development of radically different food systems which promote both affordable food and farmworker livelihood. With a range of actions available, I should be able to find a form of action that is commensurate with my specific resources and strengths. By connecting empirical experience with causal processes of oppression, critical realism builds a conceptual bridge linking structure and agency. We see how our experiences are profoundly shaped by durable social and political forces (both as agents and targets of oppression), how we participate in their daily re-creation and how we can interrupt them, individually and collectively. In this way a critical realism-informed critical consciousness can activate social change.

Implications for social work

As a profession, social work is no stranger to supporting Coloniality. It consistently promotes White women such as Mary Richmond and Jane Addams as the "founders" of the profession while ignoring the contributions made by BI-POC forerunners who, because of their race, were not valued. Also, the profession used White social and ideological structures to determine how problems should be addressed, including its reliance on individualism (Wright et al., 2021). A critical realist approach to understanding critical consciousness development allows social workers to deepen their awareness of systemic oppression and commitment to social change action. The literature indicating to what extent social workers are involved in social action, tends to focus on the bifurcation of micro and macro trained professionals. Regardless of the increase or decrease in social worker's engagement in political action, most of the findings translate into macro trained professionals reporting being more politically engaged (Choi et al., 2015; Ezell, 1993; Reeser & Epstein, 1990; Ritter, 2007; Wolk, 1981). Understanding that a major component of macro level social work is working within larger institutions and organizations whose mission is to create social change, it is not surprising that these individuals are more likely to report social activism. Mattocks, 2018) suggests macro level social workers may report higher levels of social action because it is part of their job requirement. Regardless of micro or macro orientation, research suggests that most social workers are not sufficiently prepared by their social work education program to participate in social action on the policy level (Ritter, 2007).

Many programs lack adequate training around how to intervene on a macro level (Choi et al., 2015) and the profession itself has been disinclined to provide clarity on what constitutes social justice and how it is operationalized in the field (O'Brien, 2010). The profession may be relying too heavily on the historical mission of social work, socialization into the profession by other social workers, and the implicit curriculum many programs employ as a means to educate students around important concepts and skills regarding socially just change (Bhuyan et al., 2017). Social workers may be very adept at verbalizing the profession's commitment to social justice while still lacking fundamental knowledge, skills, and techniques that endeavor to dismantle all forms of injustice. As described above, critical realist conscientization can help social workers recognize the various levels at which oppression works and explore the interplay of structure and agency in pursuit of avenues for social change. This exploration can take place in a range of social work spaces from agency board rooms to clinical supervisions to community meetings and result in strategic tangible action.

Additionally, this form of critical consciousness development can promote a sense of interconnectedness which demands accountability. As described in the example of the poultry scenario, critical consciousness can alter one's empirical experience through changing the lens with which one perceives reality. This ontological shift can potentially break down barriers that obstruct social change action. For members of targeted groups, making connections across empirical experience can lead to networks of solidarity. For example, a low wage worker in another industry having a greater risk for contracting COVID-19 because their job does not allow them to self-isolate, may recognize connections between their experiences of exploitation and different, but related experiences of the farmworker. This analysis may change their empirical perception of the inexpensive food they purchase, their need for cheap food for survival, the labor exploitation involved in generating cheap food and the Capitalistic mechanisms that bind so many in a system of exploitation.

Conclusion

The general concepts addressed in the matrix of coloniality, critical consciousness, and varying levels of reality that constitute our lived experiences are certainly not new to the field of social work. However, the three "C's (coloniality of power, critical realism and critical consciousness) can provide a shared language to communicate the ontology of oppression. The COVID-19 pandemic has exposed a variety of social forces that have increased the risk of infection, hospitalization, and death in BIPOC various communities. Providing a framework that can help explain the disparities using different layers of reality can assist in finding ways to eliminate them; thus, offering a pathway to accountability and the praxis of racial justice. It can encourage social workers to inform professional judgements by systematically making connections between their own lived experiences, the experiences of clients and communities, and social forces of power. As a collective process it can spur conversation and generate questions about social work as a field, social workers" interactions with clients, and social workers' roles as informed citizens. As a field, this form of conscientization can encourage social workers to evaluate the ways in which our professional activities may support one or all axes of the matrix of coloniality at the actual and causal levels in ways that are initially difficult to glean from our empirical experiences.

This framework can be used as a tool with clients to evaluate empirical experiences, connect to aggregate empirical patterns, examine actual systems and understand the role of structural forces. Including this analysis in individual work with clients provides a systematic strategy to identify and discuss the influence of the coloniality of power on client experiences and their own inherent human agency. At a personal level, we are accountable for being aware of how we support structures of oppression outside of direct client/community contact. In this case, the personal also connects to the professional, in that it is possible that a social worker could be advocating for a client in his/her professional role, while unknowingly perpetuating the very system that is oppressing that client.

Social work has always been a profession that works with underrepresented and historically oppressed populations and is committed to the elimination of all forms of oppression and other forms of social injustice. As a consequence, the work that social workers have done and do is often marginalized, like the populations they serve. Social work practitioners, educators, and researchers are therefore obligated to use theories that help explain the social forces that exploit power and privilege to oppress, as well as emancipatory frameworks that utilize human agency to transform oppressive structures.

Disclosure statement

No potential conflict of interest was reported by the author(s).

ORCID

Lisa Werkmeister Rozas (iD) http://orcid.org/0000-0002-7821-3490

References

Adams, M. E., Bell, L. A. E., & Griffin, P. E. (2007). *Teaching for diversity and social justice*. Routledge/Taylor & Francis Group.

Alexander, M. (2010). *The New Jim Crow: Mass Incarceration in the Age of Colorblindness*. New Press.

Allen, S. E. (2019). Doing Black Christianity: Reframing Black Church scholarship. *Sociology Compass*, 13(10), e12731. https://doi.org/10.1111/soc4.12731

Archer, M., Bhaskar, R., Collier, A., Lawson, T., & Norrie, A. (1998). *Critical realism: Essential readings*. Routledge.

Arcury, T. A., Quandt, S. A., Cravey, A. J., & Elmore, R. C. (2001). Farmworker reports of pesticide safety and sanitation in the work environment. *American Journal of Industrial Medicine*, 39(5), 487–498. https://doi.org/10.1002/ajim.1042

Arcury, T. A., Weir, M., Chen, H., Summers, P., Pelletier, L. E., Galvan, L., Bischoff, W. E., Mirabelli, M. C., & Quandt, S. A. (2012). Migrant farmworker housing regulation violations in North Carolina. *American Journal of Industrial Medicine*, 55(3), 191–204. https://doi.org/10.1002/ajim.22011

Bacon, J. M., & Norton, M. (2019). Colonial America today: US empire and the political status of native American nations. *Comparative Studies in Society and History*, 61(2), 301–331. https://doi.org/10.1017/S0010417519000069

Bhuyan, R., Bejan, R., & Jeyapal, D. (2017). Social workers' perspectives on social justice in social work education: When mainstreaming social work masks structural inequalities. *Social Work Education*, 36(4), 373–390. https://doi.org/10.1080/02615479.2017.1298741

Bibbins-Domingo, K. (2020). This time must be different: Disparities during the COVID-19 pandemic. *Annals of Internal Medicine*, 173(3), 233–234. https://doi.org/10.7326/M20-2247

Chen, Y. H., Glymour, M., Riley, A., Balmes, J., Duchowny, K., Harrison, R., & Bibbins-Domingo, K. (2021). Excess mortality associated with the COVID-19 pandemic among Californians 18–65 years of age, by occupational sector and occupation. *PLoS One*, 16(6), e0252454.

Choi, M. J., Urbanski, P., Fortune, A. E., & Rogers, C. (2015). Early career patterns for social work graduates. *Journal of Social Work Education*, 51(3), 475–493. https://doi.org/10.1080/10437797.2015.1043198

Cornish, M. (2006). Closing the global gender pay gap: Securing justice for women's work. *Comparative Labor Law and Policy Journal*, 28, 219–249.

Ezell, M. (1993). The political activity of social workers: A post-Reagan update. *Journal of Sociology and Social Welfare*, 20(4), 81–97.

Farmworker Justice. (2017). *Farmworker justice: Empowering farmworkers to improve their living and working conditions since 1981*. www.farmworkerjustice.org

Feagin, J. R. (2020). *The white racial frame: Centuries of racial framing and counter-framing*. Routledge.

Fletcher Hill, J. (2017). *The sin of white supremacy: Christianity, racism, & religious diversity in america*. Orbis Books.

Freire, P. (1973). *Education for critical consciousness*. Continuum.

Freire, P. (2018). *Pedagogy of the oppressed*. Bloomsbury publishing USA.

Garran, A. M., & Werkmeister Rozas, L. (2013). Cultural competence revisited. *Journal of Ethnic & Cultural Diversity in Social Work*, 22(2), 97–111. https://doi.org/10.1080/15313204.2013.785337

Gorski, P. (2019). *American covenant: A history of civil religion from the puritans to the present*. Princeton University Press.

Grosfoguel, R. (2011). Decolonizing post-colonial studies and paradigms of political-economy: Transmodernity, decolonial thinking, and global coloniality. *Transmodernity: Journal of Peripheral Cultural Production of the Luso-Hispanic World*, 1(1). https://doi.org/10.5070/T411000004

Grosfoguel, R. (2016). What is racism? *Journal of World-Systems Research*, 22(1), 9–15.

Hockey, N. (2010). Engaging postcolonialism. *Journal of Critical Realism*, 9(3), 353–383. https://doi.org/10.1558/jcr.v9i3.353

Houston, S. (2001). Beyond social constructionism: Critical realism and social work. *The British Journal of Social Work*, 31(6), 845–861. https://doi.org/10.1093/bjsw/31.6.845

Husain, A. (2017). Retrieving the religion in racialization: A critical review. *Sociology Compass*, 11(9), 1–8. https://doi.org/10.1111/soc4.12507

Mattocks, N. O. (2018). Social action among social work protections. *Examining the Micro-macro Divide*, 63(1), 7–16. https://doi.org/10.1093/sw/swx057

McLaughlin, J., & Hennebry, J. (2013). Pathways to precarity: Structural vulnerabilities and lived consequences for migrant farmworkers in Canada. In L. Goldring & P. Landolt, (Eds.), *Producing and Negotiating Non-citizenship: Precarious Legal Status in Canada*, University of Toronto Press, 175–194.

Morris, A. (2001). Social movements and oppositional consciousness. In J. J. Mansbridge & A. Morris (Eds.), *Oppositional consciousness: The subjective roots of social protest* (pp. 20–37). The University of Chicago Press.

Murphy, J., Samples, J., Morales, M., & Shadbeh, N. (2015). "They talk like that, but we keep working": Sexual harassment and sexual assault experience among Mexican Indigenous farmworker woman in Oregon. *Journal of Immigrant and Minority Health*, 193(1), 118–125. https://doi.org/10.1016/j.jneumeth.2010.08.011

Nelson, L. (2008). Racialized landscapes: Whiteness and the struggle over farmworker housing in Woodburn, Oregon. *Cultural Geographies*, 15(1), 41–62. https://doi.org/10.1177/1474474007085782

O'Brien, M. (2010). Social justice: Alive and well (partly) in social work practice? *International Social Work, 54*(2), 174–190. https://doi.org/10.1177/0020872810382682

Pan, D., Sze, S., Minhas, J. S., Bangash, M. N., Pareek, N., Divall, P., , and Pareek, M. (2020). The impact of ethnicity on clinical outcomes in COVID-19: A systematic review. *EClinicalMedicine, 23*(June), 100404. https://doi.org/10.1016/j.eclinm.2020.100404

Patomäki, H., & Wight, C. (2000). After postpositivism? The promises of critical realism. *International Studies Quarterly, 44*(2), 213–237. https://doi.org/10.1111/0020-8833.00156

Quandt, S. A., Wiggins, M. F., Chen, H., Bischoff, W. E., & Arcury, T. A. (2013). Heat index in migrant farmworker housing: Implications for rest and recovery from work-related heat stress. *American Journal of Public Health, 103*(8), 24–26. https://doi.org/10.2105/AJPH.2012.301135

Quijano, A. (2007). Coloniality and modernity/rationality. *Cultural Studies, 21*(2–3), 168–178. https://doi.org/10.1080/09502380601164353

Ramos, A. K. (2018). A human rights-based approach to farmworker health: An overarching framework to address the social determinants of health. *Journal of Agromedicine, 23*(1), 25–31. https://doi.org/10.1080/1059924X.2017.1384419

Reeser, L. C., & Epstein, I. (1990). *Professionalization and activism in social work: The sixties, the eighties, and the future.* Columbia University Press.

Ritter, J. (2007). Evaluating the political participation of licensed social workers in the new millennium. *Journal of Policy Practice, 6*(4), 61–78. https://doi.org/10.1300/J508v06n04_05

Robinson, E., Nguyen, H. T., Isom, S., Quandt, S. A., Grzywacz, J. G., Chen, H., & Arcury, T. A. (2011). Wages, wage violations, and pesticide safety experienced by migrant farmworkers in North Carolina. *New Solutions: A Journal of Environmental and Occupational Health Policy, 21*(2), 251–268. https://doi.org/10.2190/NS.21.2.h

Rozas, L. W., & Miller, J. (2009). Discourses for social justice education: The web of racism and the web of resistance. *Journal of Ethnic & Cultural Diversity in Social Work, 18*(1–2), 24–39.

Schoenfeld Walker, A., Singhvi, A., Holder, H., Gebeloff, R., & Avila, Y. (2021, March 5)*Pandemic's racial disparities persist in vaccine rollout.* New York Times. https://www.nytimes.com/interactive/2021/03/05/us/vaccine-racial-disparities.html

Telega, S. W., & Maloney, T. R. (2010). *Legislative actions on overtime pay and collective bargaining and their implications for farm employers in New York State, 2009-2010.* Charles H. Dyson School of Applied Economics and Management College of Agriculture and Life Sciences Cornell University

Waugh, I. M. (2010). Examining the sexual harassment experiences of Mexican immigrant farmworking women. *Violence Against Women, 16*(3), 237–261. https://doi.org/10.1177/1077801209360857

Whitehead, A. L., & Perry, S. L. (2020). *Taking America Back for God: Christian Nationalism in the United States.* Oxford University Press.

Wolk, J. L. (1981). Are social workers politically active? *Social Work, 26*(4), 283–288. https://doi.org/10.1093/sw/26.4.283

Wright, K. C., Carr, K. A., & Akkin, B. A. (2021). The whitewashing of social work history: How dismantling racism in social work education begins with an equitable history of the profession. *Advances in Social Work, 21*(2/3), 274–297. https://doi.org/10.18060/23946

Xu, S., & Li, Y. (2020). Beware of the second wave of COVID-19. *The Lancet, 395*(10233), 1321–1322. https://doi.org/10.1016/S0140-6736(20)30845-X

Anti-racism and equity-mindedness in social work field education: a systematic review

Candice C. Beasley, Melissa I. Singh and Katherine Drechsler

ABSTRACT

The purpose of this article is for schools of social work to critically evaluate their Field Education Programs for practices that impede anti-racist and equity-minded learning environments. This systematic review synthesizes the literature over the last decade to examine literature that may assist with field education evaluation. After reviewing the literature, five studies were located that addressed anti-racism within social work education; however, zero evaluative tools were located. Therefore, the authors have compiled a series of evaluative questions, as part of an initial evaluative tool, in the critical examination of anti-racist and equity-mindedness in their field education departments.

Introduction

Within social work practice and social work education, anti-racism and equity-mindedness are two terms that are discussed at great length. With anti-Black and anti-Asian racism becoming the norm and not the exception, the profession of social work is now forced to transition these discussions of anti-racism and equity-mindedness into one of action, but how? Unlike the immediate transition schools of social work were forced to make at the start of the COVID-19 pandemic, transitioning racist and oppressive systems, such as the academy, into one that embraces anti-racism and equity will be a more difficult task to pursue. Nonetheless, as social work attempts to stay relevant and compete within this nuanced technological age (i.e., online schools of social work, telehealth, etc.), now more than ever, social work education and the profession are being connected to increasing and varied Black, Indigenous, and People of Color (BIPOC) communities. As this connection increases, both social work students and professionals will be more engaged with Black, Brown, and Asian communities, the very communities currently being harmed by anti-Black and anti-Asian racism.

Schools of social work must be mindful that students entering the academy are part of these communities. Because of this, social work students are demanding change within the academy, change within social work, and demanding equitable opportunities within social work field education. Schools of social work are now being called from rhetorical conversations regarding anti-racism and equity-mindedness to implementing an anti-racist curriculum and equitable field education experiences. The only way that schools of social work can begin this implementation is through authentic introspection and critical analysis of both its curriculum and its field education program.

Due to field education's branding as the signature pedagogy of social work education, the authors will focus upon the field education component relevant to schools of social work and all social work entities that are dependent upon the Council of Social Work Education (CSWE) for accreditation.

Hopefully, all schools of social work, along with their administration, will utilize the guided self-evaluative tool, provided within this paper, to critically analyze their field education policies and processes for embedded racist and inequitable practices so that social work education may become genuinely anti-racist and equity-minded.

Diversity, equity-mindedness and anti-racism

For social work education to be equity-minded and anti-racist, it is to embrace the meaning of diversity, as set forth by the social work profession, while requiring a brave learning environment or space to connect pedagogy to content. NASW's Cultural Competence Standards (National Association of Social Workers, 2015) defines diversity as a concept that is "more than race and ethnicity, it includes the sociocultural experiences of people inclusive of, but not limited to, national origin, color, social class, religious and spiritual beliefs, immigration status, sexual orientation, gender identity or expression, age, marital status, and physical or mental disabilities" (pg.9).

According to Arao and Clemens (2013), brave spaces encourage leaning into a conversation with diverse perspectives to dialogue on topics such as social justice authentically. "The term 'equity-mindedness' refers to the perspective or mode of thinking exhibited by practitioners who call attention to patterns of inequity in student outcomes. These practitioners are willing to take personal and institutional responsibility for their students' success and critically reassess their practices. It also requires that practitioners are race-conscious and aware of the social and historical context of exclusionary practices in American Higher Education" (Center for Urban Education, 2021, para. 1). According to the Council on Social Work Education (2020a), equity-minded competence is achieved when an individual:

> Becomes aware of racial identity, uses disaggregated data to identify inequitable racial and ethnic outcomes, reflects on racial and ethnic consequences of practices, exercises agency to produce racial and ethnic equity, views the classroom as a racialized space and self-monitor interactions with students of color(para. 1).

In staying relevant to anti-racist efforts, social work educators need to be abreast of emerging research. "If we decry structural racism but return to the behaviors and processes that led us to this moment, this inexcusable stagnation will continue" (Barber et al., 2020, p. 1440). Stagnation allows for the perpetuation of racist and oppressive policies and actions. Identifying the importance and acknowledging and honoring the lived experiences of social work students is just as crucial as anti-racist efforts made in the communities social workers serve. The lived experiences of social work students become especially relevant during field education.

National Association of Social Work (NASW)

The need for the social work profession and social work education, including social work field education, to be anti-racist and equity-minded is well documented. Racial justice is the heart of social work practice, and many individuals choose the social work profession because of the social justice values and principles embedded in the profession's Code of Ethics. Social workers have the responsibility to recognize that structural racism impacts their well-being and their professional life. The NASW Code of Ethics, first approved on October 13, 1960, sets forth standards to guide the social work profession, including guiding values and principles. The NASW preamble states:

> The primary mission of the social work profession is to enhance human well-being and help meet the basic human needs of all people, with particular empowerment of people who are vulnerable, oppressed, and living in poverty. A historic and defining feature of social work is the profession's focus on the individual well-being in a social context of well-being of society (National Association of Social Workers, 2021, p. 1).

In 2008, the NASW Delegate Assembly revised the Code of Ethics and developed the Ethical Standard 1.05 Cultural Awareness and Social Diversity. In 2017 a revision to the NASW Code of Ethics changed the title of this standard to Cultural Competence and Social Diversity (National Association of Social Workers, 2017a).

The 2021 amendment validated that this change diluted the commitment and spirit of this standard and reinstated the cultural competence language to support the focus on cultural competence frameworks and approaches. The 2021 amendments to the Code of Ethics revised Standard 1.05 Culture Competence to include language validating the importance of social workers to demonstrate the understanding of culture and act against oppression, racism, discrimination, and inequalities, and acknowledge personal privilege. This same standard was further amended to include that social workers should demonstrate awareness and cultural humility by engaging in critical self-reflection, recognizing that clients are the experts of their own culture, committing to lifelong learning, and holding institutions accountable for advancing cultural humility. The Code of Ethics remains the current standard that empowers social workers to advocate for the oppressed and recently was amended to include additional value statements to highlight the importance of cultural competence in the social work profession (National Association of Social Workers, 2021).

Almost 30 years ago, NASW launched an initiative to challenge racism at the individual, organizational, and societal levels, "Color in White Society" (White, 1982). The social work profession currently is over 70% white (Council on Social Work Education, 2015a) and functions in a system of white supremacy. In 2017, NASW aimed to address the impact of institutional racism in A Call to Action for Institutional Racism & the Social Work Profession which challenged the social work profession to address structural racism, both in limiting its negative influences and creating outcomes to address it (National Association of Social Workers, 2017b).

In addition to the Code of Ethics Standard 1.05, in 1966, NASW further defined standards that required social workers to demonstrate cultural competence by adopting the first Standards for Cultural Competence (National Association of Social Workers, 2000). These standards were revised in 2001 (National Association of Social Workers, 2001) and 2015 as the Standards and Indicators for Cultural Competence in Social Work Practice (National Association of Social Workers, 2015). These standards affirmed that the social work profession and social workers have an "ethical responsibility to be culturally competent" (National Association of Social Workers, 2001, p. 7).

Council on Social Work Education

The Council on Social Work Education (CSWE) was formed in 1952 and continues to be the accrediting body for social work education. CSWE declares that:

> Accreditation is a system for recognizing educational institutions and professional programs affiliated with those institutions as having a level of performance, integrity, and quality that entitles them to the confidence of the educational community and the public they serve(Council on Social Work Education, 2015b, p. 4).

CSWE uses the Educational Policy and Accreditation Standards (EPAS) to accredit baccalaureate and master's level social work education programs. Educational Policy 2.1.4 requires that explicit and implicit curriculum, through intentional design, achieves the ability of social work students to engage in diversity and difference in practice. CSWE, through its policy curriculum statement, provides a mandate for infusion of multicultural content in academic courses (Julia, 2000) identifying it as essential. A required expectation of accredited schools of social work is to also achieve cultural diversity in enrollment of students and hiring of faculty (McMahon & Allen-Meares, 1992).

The pedagogy involved in integrating diversity and difference in practice within social work curriculums is not explicit. Emphasizing diversity in the content of social work education has not been a priority in the history of social work education. As far back as 2004, Fong and Lum indicated, "as social work educators, policymakers, and researchers, we have been successful in identifying the need for cultural competence practice; however, we have not been effective enough in facilitating the process for infusing cultural competence throughout the social work curriculum" (Fong & Lum, 2004, p. 19). Garran and Werkmeister Rozas (2013) identified the need for a more comprehensive plan for schools of social work for cultural competence that includes intersectionality of race in schools of social work and social work organizations.

Grand Challenges for Social Work

The American Academy of Social Work and Social Welfare (2020), initiated the Grand Challenges for Social Work and the Grand Challenge to Eliminate Racism on June 26, 2020. This challengepromotes culturally grounded prevention and intervention programs to eradicate racist policies, bias, and discriminatory practices of inequality (Teasley et al., 2021). Social work education is committed to abolishing racism through research, teaching, and work within communities. Social justice is the core social work principle of this challenge. The social work field, including social work education, is challenged to embrace a more explicit and continued commitment to its role in eliminating systemic racism (Rao et al., 2021). It is not only social work education that is to be attentive to this Grand Challenge. The profession must be more attentive to the intent of the Grand Challenges of Social Work by: (1) understanding and internalizing the intent of cultural competence, (2) being mindful of and correcting performative displays of cultural competence and (3) engaging in self-reflective practices that will assist in understanding the tenants of cultural competence while actively engaging in corrective action plans that will reduce values, beliefs and practices that undermine the embracing of cultural competence.

Field Education

CSWE requires field education as part of social work education. Field education's philosophy was originally based on social work students learning to practice the profession through experiences in an agency supervised by social work practitioners. Active involvement in real professional experiences serves to prepare students for the fundamental duties of the social work profession. The 2008 CSWE EPAS identified the field education experience as the signature pedagogy of the social work profession. "Signature pedagogy is a central form of instruction and learning to socialize students to perform the role of practitioner–it contains pedagogical norms with which to connect and integrate theory and practice" (Council on Social Work Education, 2008, p. 8). The term signature pedagogy originated in 2005, by Lee Schulman, which defines how educational approaches teach students fundamental elements of the profession (Shulman, 2005). Experiences in field education build many competencies in social work students. For example, the view of holistic competence is the integration of field education experiences and critical thinking while applying knowledge learned in course curriculum to practice (Bogo, 2018). Arguably then, the importance of examining how the field education experience is anti-racist and equity-minded is essential in preparing social work students for the necessary competence in engaging in diversity and difference in practice.

Why the Call to Action?

With the emergence of anti-Black and anti-Asian racism, coupled with the staggering health disparities within BIPOC communities stemming from COVID-19, schools of social work are compelled to revisit if the current oppressive systems in which they function are still plausible. Currently, anti-racist and equity-minded approaches, based upon Critical Race Theory (CRT), provide the tools to impact immediate and meaningful change within social work education. However, the integration of anti-racist and equity-minded practices, within social work education, will not be effortless due to the multifaceted tiers within the academy supported by racist and oppressive values and practices. Therefore, if we are to begin this implementation in a meaningful way, social work programs, as a collective, must analyze: (1) the structure of racism within their institutions and administration, (2) pedagogical models that incorporate implicit and explicit themes of oppression, (3) accountability of faculty in modeling anti-racist and equity-minded practices, and (4) the curriculum and educational experience of social work students.

While analyzing racism, we are called to alter how privileged groups view racism and oppression; however, it is also the responsibility of schools of social work to analyze how systems and institutions reproduce privilege (Yee, 2016). This reproduction may be strengthened by administrators and faculty who emerge from privileged groups as there are roles that they must ascribe to as it relates to oppressive systems within institutions, which further reinforces the avoidance of engaging in the dismantling of anti-oppressive practices, processes, and structures (Yee, 2016) within schools of social work. Pedagogical models that sustain racist and oppressive beliefs, values, and practices must also be considered as such models undermine the mission and values of social work as they cause direct and often irreparable harm to BIPOC social work faculty and students. To assist with antiquated and oppressive pedagogical models, Wagaman et al. (2019) identify that transformational models of teaching and learning are needed to prepare students to engage in discourse surrounding racism and address racism in practice.

Further, both the curriculum and the classroom must provide a safe space for students to dialogue and have meaningful discourse regarding racism and oppression. Although safe space does not equate to comfortability, social work students require this dialogue to build competencies in becoming meaningful change agents who oppose social injustice. The incorporation of anti-racist and equity-minded practice is also positively correlated with social work students' educational and social experience. Swick et al. (2021) purport that safe spaces and affinity groups are needed in facilitating meaningful dialogue among faculty and students as silence among white faculty and white classmates is frustrating and concerning to BIPOC students (Swick et al., 2021); hence, negatively altering the educational and social experience of the BIPOC student. Analysis of the pedagogical models used within the social work curriculum is not managed as a preemptive task; instead, it is to be a collective, purposeful and informed analysis so that "next steps" stemming from the analysis will allow faculty members and students to engage in this discourse while discouraging performative allyship by white counterparts (Swick et al., 2021).

Schools of social work are also to be attentive to the nuanced characteristics of students entering into the school. With the emergence of millennial social work students, the literature finds these students more racially tolerant (Wagaman et al., 2019). Nonetheless, these students continue to be resistant to and uncomfortable with discussing issues of race and racism experienced through feelings of discomfort, guilt, shame, and anxiety (Hamilton-Mason & Schneider, 2018; Wagaman et al., 2019). Because of this, social work education must revisit how values and beliefs, surrounding racism, are addressed within the classroom and the field education experience. More importantly, schools of social work are responsible for ensuring that all faculty members are comfortable and competent in discussing race, racism, and oppression beyond mere theoretical course instruction. Social work programs must be deliberate in considering this call to action to honor the mission and values that govern our education (CSWE) and profession (NASW) while remaining relevant to competing social science educational programs.

Methodology

For this paper, a systematic review was completed using the PRISMA Flow Diagram (Page et al., 2021). This ten-year literature review attempted to locate a concise body of literature, or evaluative tool, which guides schools of social work in critically evaluating their field education departments for practices that may undermine both anti-racist and equity-minded learning environments.

Inclusionary Articles	Peer Reviewed	MSW Education	Field Education	Published in USA	Anti-racism
Feize & Gonzalez (2018)	X	X	-	X	X
Hamilton-Mason & Schneider (2018)	X	X	X	X	X
Swick, Dyson &Webb (2021)	X	X	X	X	X
Wagaman, Odera & Fraser (2019)	X	X	X	X	X
Yee (2016)	X	X	-	X	X

Figure 1. Systematic Review Results

PRISMA Flow Diagram

The authors completed the following steps to gather generalized, operational, and current literature regarding guidance in critically evaluating social work field education departments for anti-racist and equity-minded practices. To locate this body of literature, a systematic search was performed utilizing the following EBSCO Research databases: Academic Search Complete, APA PsycArticles, APA PsycInfo, Educational Administration abstracts, E-Journals, Professional Development Collection, Psychology & Behavioral Sciences Collection, Race Relations Abstracts, Social Work Abstracts, SocINDEX w/ Full Text. Through a 10 year review, these databases were searched from April 2011 through April 2021, screening for "peer-reviewed" publications. The keywords utilized were: "Masters of social work programs," "MSW Programs," "Anti-Racism," "Social Work Education," "Field Education," "field placement," and "practicum." Inclusionary criteria entailed that articles were to be published within the last ten years, the articles must be peer-reviewed, in the English language with full text available. The articles also were to be specific to MSW Programs located within the United States of America and identify anti-racism and/or anti-racist practices. Exclusionary criteria included: articles and journals that were not peer-reviewed, articles published prior to the last ten years, articles that were not published within the United States of America, were not in the English language, articles that did not include mention of "anti-racism" and "field education," and articles that did not include MSW programs.

Results

Through the use of the 2020 PRISMA Flow Diagram (Page et al., 2021), records identified, through the keyword search, yielded 872 possible articles with 50 articles specifically utilizing the concept of "social work" and "field education." Therefore, fifty full-text articles were assessed for eligibility. Out of the 50 articles, 45 articles failed to meet the inclusionary standards (Figure 2). A total of five studies (n = 5) met all aspects of the inclusionary criteria; however, 0 articles provided an anti-racist evaluative tool that may be used for social work field education programs (Figure 1).

Discussion

The articles identified within this systematic review collectively provide both points to consider and guidance on how social work educators can begin conceptualizing and implementing anti-racist and equity-minded practice within social work education. These articles identify the need for analysis and research within social work education infrastructures. The research of Yee (2016), explores racism and oppression in school structures and curriculum. Through the research of Feize and Gonzalez (2018), Hamilton-Mason and Schneider (2018), and Wagaman et al. (2019), antiquated pedagogical models embedded within social work education are discussed. Finally, Swick et al. (2021) address the social work students' educational and social experience while in social work programs.

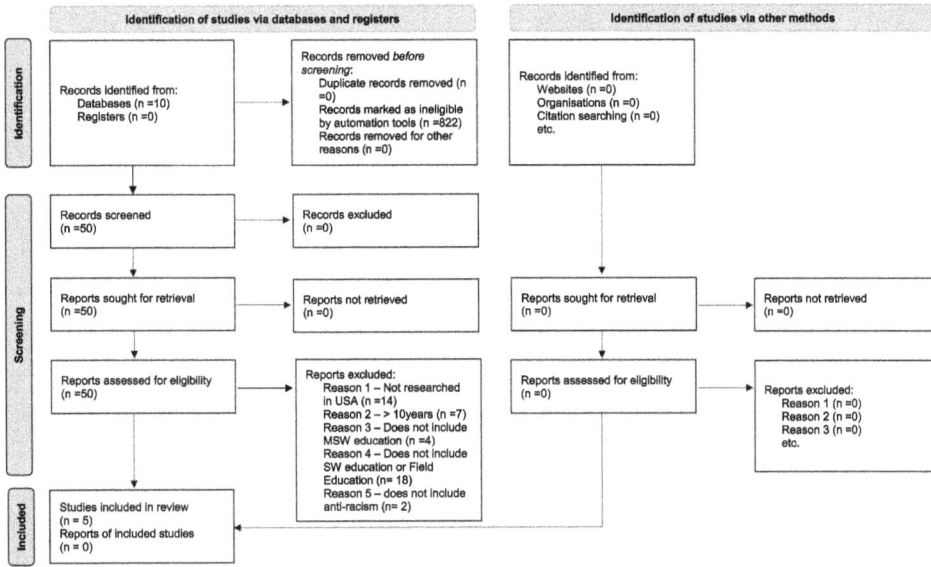

Figure 2. PRISMA Flow Diagram Results

When analyzing school structures and social work curriculums, we consider that BIPOC faculty and students experience racism within their daily lives. Frequently, these groups feel "unheard" within the academy as structural implications of racism fail to be examined (Yee, 2016). Further, changing how privileged groups view racism and oppression is imperative; however, Yee (2016) further recognizes that although, as social workers, we are taught about anti-oppressive perspectives and values, our scope of action is limited by the reality of the discriminatory and exclusionary institutional structures of the academy. While considering equity-minded practice, it seems as though social work school administrators are hesitant in implementing this concept. Perhaps there is concern that creating an equity-minded space means lowering expectations and standards. On the contrary, equity-minded practices focus upon developing both sensitive and supportive mechanisms that will assist the creation of pathways leading to the promotion of access and participation (Yee, 2016). The literature further identifies that in an attempt to assuage "white guilt," social work education tends to practice "safe" education as to negate controversial dialogue within the classroom (Yee, 2016); yet, this is counterproductive to what NASW and CSWE call us to do, which may inadvertently cause schools of social work to undermine its success in adhering to accreditation standards and the ethical considerations of social work practice.

Analysis of antiquated pedagogical models and course instruction is also needed to remove oppressive and discriminatory aspects of social work education. Social work students need educational and field experiences that will allow them to process race, racism, and social injustice sans fear of grade penalty or perception of diminished professional competence (Wagaman et al., 2019). Hamilton-Mason and Schneider (2018) identify themes that may be considered as critical components to anti-racist pedagogy. These themes include incorporating activity-based learning, sharing perspective and learning from peers, addressing white privilege beyond mere course readings, allowing for emotional discomfort, and being exposed to "new language" in discussing racism. If an anti-racist and equity-minded implementation is successful, active participation from social work faculty is needed. The literature reveals the importance of the faculty role in creating an anti-racist space for social work students. For example, the research of Feize and Gonzalez (2018) found that some faculty members, within their study, believed that teaching anti-racist tenants is unnecessary and expect that issues of racism will diminish once minorities are embraced within society. Further, even when BIPOC faculty were interviewed, Feize and Gonzalez (2018) found that concepts of white and capitalist relationships

Level 1 – Field Placement Process

- Is their verbiage embedded in the school's memorandum of understanding, with the field sites, regarding the need for field experiences that are anti-racist and equity-minded?
- Is the field placement process transparent? Are the students provided field placement choices?
- As it relates to race, ethnicity, and culture, are contracted field sites and partnerships with community agencies reflective of the school's student body? If so, is this reflection genuinely unbiased?
- Are field placement opportunities published and accessible to incoming and continuing students?
- Are field education opportunities that offer paid stipends and/or allow for job placement opportunities published and accessible to both incoming and continuing students?
- Are school-sponsored field education scholarships/grants/stipends available? If so, is the application and eligibility criteria published and accessible to both incoming and continuing students?
- Is there a platform or safe space available where students who are not from dominant cultures, educated about field placement opportunities that provide more financially lucrative skill sets upon graduation?
- Is the school and field education department utilizing field placements known to support and promote anti-racist and equity-minded practices within the community?
- Has the implementation of anti-racist field assignments/projects been considered? If so, what are the plans for implementation?
- Is there a review on how cultural competency is taught within field education (theories/ concepts, align field experiences with these updated concepts, anti-essentialism)?
- What outcome is the school using to identify anti-racist and equity-minded practices in the field education process?
- Other than field evaluations, are there subsequent measures in place in accessing the skill sets of diverse students?

Level 2 – School of Social Work Administration, Agency Administration, Field Leadership, Faculty, Field Instructors, Liaisons, and Preceptors

- Are field directors making intentional efforts to partner with agencies and external field instructors that reflect the community that the school's students serve? If so, can these efforts be identified?
- Have members of field faculty exemplified active participation in developing their positionality and self-awareness to understand their privilege?
- What is the process in evaluating if faculty and external field instructors inherently support anti-racist and equity-minded practice?
- Is critical multiculturalism (examining whiteness) being used among field instructors/faculty field liaisons in schools of social work that are housed in primarily white institutions (PWIs)?
- Are there opportunities for students to formally evaluate their field placement and external field instructors?
- What mechanisms are in place for faculty, field liaisons, external field instructors, and field placements that refuse to support anti-racist and equity-minded tenants and practices?
- What orientation and ongoing training opportunities are provided for field directors, faculty field liaisons, and external field instructors on the importance of anti-racist and equity-mindedness in field education?
- How has the school's administration and faculty collaborated with subsequent schools of social work in determining and establishing best practices in teaching?
- Are social work faculty members comfortable in challenging racist thoughts/ideas expressed within the classroom? If not, how will this be addressed?

(Continued)

(Continued).

Level 3 – Social Work Curriculum

- How are concepts such as cultural humility, critical reflexivity, self-awareness, and self-regulation defined and operationalized within course discussions?
- Are value-based practices embedded in the curriculum?
- Is there an identified method in connecting anti-racist perspectives within social work research courses?
- Are students taught to identify racial bias within social work research?
- As it relates to the school of social work structure by which the field education department is housed, has hegemonic masculinity and neo-colonial practices been addressed?
- Does the school's social work curriculum focus on the practice skill set needed to engage with diversity and difference in social work practice?
- Are there field courses, such as field seminars, structured with curriculum, activities, and assignments to assist students with the application of anti-racist concepts?
- As it relates to race and ethnicity, does the ratio of faculty mirror that of the student body?
- Have all faculty and school administration members identified implicit and explicit biases embedded within the school's infrastructure? Curriculum? If so, has a plan of action to address the identified issues been composed?
- Has a core course/elective been implemented, into the curriculum, that solely focuses on racism, racism in social work, and anti-racist interventions/practices?
- Have affinity groups that address race, racism, and white privilege been established for White students? What plan of action has been implemented to ensure that White faculty is primarily facilitating these groups?
- Is there an equitable and transparent process in assigning faculty members to uncompensated activities/projects/committees within the school?
- Are BIPOC faculty members continuously utilized for uncompensated activities/projects/committees within the school? If so, what plan has been implemented in providing compensation?
- What mechanisms are in place for administrators, tenured faculty, non-tenured faculty, and adjunct faculty members that refuse to support anti-racist and equity-minded tenants and practices? If harm stemming from the aforementioned is reported, how will corrective actions be implemented?

Level 4 – Social Work Students

- Are there mechanisms of student governance in place which specifically focus upon field education? If so, as it relates to race and ethnicity, do student representatives mirror the student body?
- Do social work students have active participation in choosing their field placement and their field education experience design?
- When social work students verbalize the field experience they desire, does the field education department actively pursue placements supporting the student's wishes? Are the efforts documented and reviewed with the student?
- Are field activities, experiences, and reflections assisting students with active self-awareness opportunities?
- Has the school collected a body of literature that discusses anti-racism and equity-minded concepts that are current and accessible to students?
- How are graduate students socialized to equity, diversity, and inclusion related to the social work profession?
- Are there opportunities for students to participate in anti-racist and equity-minded workshops/trainings? If so, have opportunities been identified that are free of cost? Is the school willing to sponsor students that are unable to afford the cost of these opportunities?
- What mechanisms are in place in documenting racial harm reported by students stemming from the school? Once reported, how will corrective measures be implemented, documented, and utilized for future systemic evaluation?
- Does the social work program discuss the connection of racial harm social work students may experience in the field education experience, and the profession, to self-care practices?
- What mechanisms are in place for students that refuse to support anti-racist and equity-minded practices? How will corrective actions be implemented?
- What mechanisms are in place for white students that harm BIPOC faculty and staff through the use of power and privilege? How will corrective actions be implemented?

with other nations and cultures were not addressed within course instruction. Findings also reflect that there are no "step-by-step" processes available in implementing anti-racist course instruction and dialogue (Feize & Gonzalez, 2018); therefore, making it challenging to facilitate anti-racist and equity-minded practice within the field and across course instruction.

Finally, the educational and social experience of the social work student must be considered. Swick et al. (2021) provide student-focused considerations related to anti-racist and equity-minded practices by identifying the competency needed by social work faculty. In addition, Swick et al. (2021) remind us that it is imperative that schools of social work consider the interconnection between students' social identity and how they internalize, experience, and process social injustices (Swick et al., 2021). Further, if anti-racist and equity-minded principles are to be genuinely embedded within schools of social work, faculty must actively model anti-racist approaches and pledge to lifelong personal and professional anti-racist learning (Swick et al., 2021).

NASW, CSWE, and the Grand Challenge to Eliminate Racism acknowledge the importance of diversity and culturally competent practice in the social work profession and social work education, including the field education experience. CSWE's comprehensive policy statements validate the importance that schools of social work, include content on diversity in their explicit and implicit curriculum. The Grand Challenge charges social work education to commit to anti-racist and equity-minded curriculum and field education experiences. Although the need is well documented, there is minimal literature to review how schools of social work have integrated diversity and difference content in the curriculum and field education experience. Social work can claim that it is the mission to help the oppressed and combat racism; yet, racism in our society and in the social work profession still exists. Individual social workers are responsible for promoting anti-racism and equity-minded practice, which begins with social work education, including during the field experience, which has been coined the signature pedagogy of social work education.

Embedding equity into the structures, norms, policies, and practices of field education is crucial in the call to action for anti-racism and equity-mindedness in social work field education. Universities and colleges are the more prominent institutions in which schools of social work function and thus have a significant role in developing anti-racist and equity-minded practices. The Center for Urban Education (CUE) has developed an *Equity Scorecard* to support campuses in developing equity into their structures, norms, policies, and practices. This Equity Scorecard provides theory-based tools, activities, and processes to support campuses to make equity for racial and ethnic groups in universities and colleges a goal by bridging the gap between data and the action that needs to be taken on campuses. Individuals on campus are viewed as agents and responsible for this change. Campus participants learn how to reframe racial inequities once identified on campus and individually and collectively feel empowered to take action to remedy these actions. When given the correct tools, the conversations about racism shift to equity-minded action, which allows the transformation of the teaching and learning environment to be more supportive of historically marginalized students (Felix et al, 2015). Schools of Social work should mirror this framework in developing a field education equity scorecard process to engage schools of social work in evaluating and developing anti-racist and equity-minded practices in the field education process.

Jani et al. (2016) suggest that "it may be useful to enhance training of field instructors to assist them in evaluation of their own cultural competence and student's acquisition of appropriate skills" (p. 322). Training Field Instructors on the dynamics of cultural competence and how to measure the identified indicators of cultural competence on the field placement assessment tool would be critical in gathering accurate data on the competency of engaging in diversity and difference of graduate students. The accuracy of this data is essential in analyzing, developing, and enhancing curriculum and field education policies and processes. Although individual schools of social work may be diligently assessing their field education departments in incorporating anti-racist and equity-minded practices, a review of these results infers that within the past 10 years, American social work education has not been attentive to this incorporation as a collective. The results of this systematic review provided critical information that may be utilized for social work practice.

Social Work Field Education Self-evaluation Tool

Although not well researched and therefore not prominent in peer-reviewed literature, schools of social work across the nation have implicitly and/or explicitly supported behaviors and practices that solidify racism and oppression across all pedagogical aspects of social work education. This over-arching theme is part of the "tacit knowledge" that resides among social work educators. One of the major sub-themes identified is that antiquated social work curriculums either directly or indirectly: harm BIPOC students and faculty members, are not sensitive to the needs of BIPOC students and faculty, social work faculty either mishandles or are fearful of providing safe-spaces in having difficult conversations about race, racism and privilege, and/or field education experiences that fail to support BIPOC communities through the continued use of racist and oppressive assessments and interventions. Because of this, as reported in their 2019–2020 Annual Report, CSWE's Council on Racial, Ethnic and Cultural Diversity (CRECD) has launched the Legacy Project. The purpose of this project is to collect, identify, and address themes of harm experienced by BIPOC students, harm stemming directly from schools of social work, and the students' MSW education (Council on Social Work Education, 2020b).

To further assess social work students' educational experience, schools of social work must ensure that social work administration, faculty, field education directors/liaisons/external supervisors, and field placement agencies actively engage in anti-racist and equity-minded practices. Within this era of a severe pandemic of racism, NASW's Ethics, Racism, and Inclusion Department was compelled to release a racism statement that identified and addressed active racism within the social work profession. NASW's statement reminded the profession that social work is "not to tolerate social workers who practice, condone, facilitate or collaborate with any form of discrimination or racism" (National Association of Social Workers, 2020, para. 8). Within this same statement, NASW calls all social workers to become educated regarding race and racism, examine one's behaviors related to bias and racism, and be intentional in learning about diversity (National Association of Social Workers, 2020). This statement is extended to actively evaluating all social workers and social work partners that influence social work education. It is incumbent upon schools of social work to evaluate and identify all social work and nonsocial work entities for participation in practices that undermine anti-racist and equity-minded themes whilst also appropriately addressing the severance from said entities, if needed. Further, because of this statement, social work faculty must possess and exemplify competency in facilitating difficult conversations across pedagogical venues because of this statement. Therefore, schools of social work must evaluate how faculty will be held accountable for implementing harmful course instruction, trainings, or initiatives under the auspices of "anti-racist" work.

Questions included in the following Social Work Field Education Self-Evaluation Tool have been extracted from information gathered through the systematic review, themes that have emerged from social work students and social work field departments, and implications posed by CSWE and NASW. This self-evaluation tool will be divided into four sections: Field Placement Processes, School of Social Work Administration, Agency Administration, Field Leadership, Faculty, Field Instructors, Liaisons, Preceptors; Social Work Curriculum, and Social Work Students.

Conclusion

The findings provide preliminary support that schools of social work, within the United States, have not participated in a collective effort to ensure that social work education is partaking in anti-racist and equity-minded practices. Due to the critical importance of anti-racism and equity-mindedness in the field education experience, continued research and best practices for field education are paramount. The authors hope that once schools of social work honestly examine their curriculum and field education departments, individual schools may begin to identify, triage, and address their deficiencies so that emerging social work professionals may be able to implement anti-racist and equity-minded concepts within their practice successfully, which will in-turn lead both the academy and the profession in embodying the competencies and ethics set forth by both NASW, CSWE and the vision of the Grand Challenge to Eliminate Racism.

Disclosure statement

No potential conflict of interest was reported by the author(s).

References

American Academy of Social Work and Social Welfare. (2020). *Grand challenges of social work.* https://grandchallen gesforsocialwork.org/eliminate-racism/
Arao, B., & Clemens, K. (2013). From safe spaces to brave spaces. In *The art of effective facilitation: Reflections from social justice educators* (Stylus) (pp. 135–150).
Barber, P. H., Hayes, T. B., Johnson, T. L., & Márquez-Magaña, L. (2020). Systemic racism in higher education. *American Association for the Advancement of Science, 369*(6509), 1440–1441. https://doi.org/10.1126/SCIENCE.ABD7140
Bogo, M. (2018). *Social work practice: Integrating concepts, processes, and skills* (2nd ed.). Columbia University Press.
Council on Social Work Education. (2008). *Educational policy and accreditation standards.* https://www.cswe.org/Accreditation/Standards-and-Policies/2008-EPAS.aspx
Council on Social Work Education. (2015a). *Annual statistics on Social Work Education in the United States.* https://www.cswe.org/getattachment/992f629c-57cf-4a74-8201-1db7a6fa4667/2015-Statistics-on-Social-Work-Education.aspx
Council on Social Work Education. (2015b). *Educational policy and accreditation standards.* https://www.cswe.org/Accreditation/Standards-and-Policies/2015-EPAS.aspx
Council on Social Work Education (2020a). *What is equity minded competence?* https://www.cswe.org/getattachment/Centers-Initiatives/Centers/Center-for-Diversity/Educator-Resource/February-2020/1-What-is-equity-minded-com petence.pdf.aspx
Council on Social Work Education. (2020b). *Annual report 2019-2020.* https://www.cswe.org/getattachment/About-CSWE/CSWE-Annual-Reports/CSWE_Annual-Report_19-20.pdf.aspx
Feize, L., & Gonzalez, J. (2018). A model of cultural competency in social work as seen through the lens of self-awareness. *Social Work Education, 17*(4), 472–489. https://doi.org/10.1080/02615479.2017.1423049
Felix, E. R., Bensimon, E. M., Hanson, D., Gray, J., & Klingsmith, L. (2015). Developing Agency for equity-minded change. *New Directions for Community Colleges, 2015*(172), 25–42. https://doi.org/10.1002/cc.20161
Fong, R., & Lum, D. (2004). Developing an integrated model of cultural competence in social work education. In *Education for multicultural social work practice: Critical viewpoints and future directions* (Council on Social Work Education) (pp. 19–30).
Garran, A., & Werkmeister Rozas, L. (2013). Cultural competence revisited. *Journal of Ethnic & Cultural Diversity in Social Work Education, 22*(2), 97. https://doi.org/10.1080/15313204.2013.785337
Hamilton-Mason, J., & Schneider, S. (2018). Antiracism expanding social work education: A qualitative analysis of the undoing racism workshop experience. *Journal of Social Work Education, 54*(2), 337–348. https://doi.org/10.1080/10437797.2017.1404518
Jani, J. S., Osteen, P., & Shipe, S. (2016). Cultural competence and social work education: Moving toward assessment of practice behaviors. *Journal of Social Work Education, 52*(3), 311–324. https://doi.org/10.1080/10437797.2016.1174634
Julia, M. (2000). Student perception of culture: An integral part of social work practice. *International Journal of Cultural Relations, 24*(2), 279–289. http://dx.doi.org/10.1016/S0147-1767(99)00036-X
McMahon, A., & Allen-Meares, P. (1992). Is social work racist? A content analysis of recent literature. *Social Work, 37* (6) , 533–539.
National Association of Social Workers. (2000). *Code of ethics.* NASW.
National Association of Social Workers. (2001). *Standards for cultural competence in social work practice.* NASW.
National Association of Social Workers. (2015). *Standards and indicators for cultural competence in social work practice.* NASW.
National Association of Social Workers. (2017a). *Code of ethics.* https://www.socialworkers.org/About/Ethics/Code-of-Ethics/Code-of-Ethics-English
National Association of Social Workers (2017b). *Institutional racism in the Social Work profession: A call to action.* https://www.socialworkers.org/LinkClick.aspx?fileticket=SWK1aR53FAk%3D&portalid=0
National Association of Social Workers. (2020). *An important message about racism.* https://www.socialworkers.org/About/Ethics/Ethics-Education-and-Resources/Ethics-Resources-for-Racial-Equity/A-Message-About-Racism
National Association of Social Workers (2021). *Code of ethics.* https://www.socialworkers.org/About/Ethics/Code-of-Ethics/Code-of-Ethics-English

Page, M. J., McKenzie, J. E., Bossuyt, P. M., Boutron, I., Hoffman, T. C., Muldrow, C. D., Shamseer, L., Tetzlaff, J. M., Akl, E. A., Brennan, S. E., Chou, R., Glanville, J., Grimshaw, J. M., Hróbjartsson, A., Lalu, M. M., Li, T., Loder, E. W., Mayo-Wilson, E., McDonald, S., McGuinness, L. A., Stewart, L. A., Thomas, J., Tricco, A. C., Welch, V. A., Whiting, P., & Moher, D. (2021). The PRISMA 2020 statement: An updated guideline for reporting systematic reviews. *Systematic Reviews*, 10(1), 89. https://doi.org/10.1186/s13643-021-01626-4

Rao, S., Woo, B., Maglalang, D. D., Bartholomew, M., Cano, M., Harris, A., & Tucker, T. B. (2021). Race and ethnicity in the social work grand challenges. *Social Work*, *66*(1), 9–17. https://doi.org/10.1093/sw/swaa053

Shulman, L. S. (2005). Signature pedagogies in professionals. *Daedalus*, *134*(3), 52–59. https://doi.org/10.1162/0011526054622015

Swick, D. C., Dyson, Y. D., & Webb, E. (2021). Navigating a pandemic, racial disparities, and social work education through the lens of the NASW Code of Ethics. *Reflections: Narratives of Professional Helping*, 27(1), 84–93.

Teasley, M. L., McCarter, S., Woo, B., Conner, L. R., Spencer, M. S., & Green, T. (2021).*Grand challenges for social work initiative*. https://grandchallengesforsocialwork.org/wp-content/uploads/2021/05/Eliminate-Racism-Concept-Paper.pdf

Wagaman, M. A., Odera, S. G., & Fraser, D. V. (2019). A pedagogical model for teaching racial justice in social work education. *Journal of Social Work Education*, American Academy of Social Work & Social Welfar *55*(2), 351–362. https://doi.org/10.1080/10437797.2018.1513878

White, B. W. (Ed.). (1982). *Color in a white society*. National Association of Social Workers.

Yee, J. Y. (2016). A paradox of social change: How the quest for liberation reproduces dominance in higher education and the field of social work. *Social Work Education*, *35*(5), 495–505. https://doi.org/10.1080/02615479.2016.1170113

"Talking about race is exhausting": social work educators' experiences teaching about race and racism

Ebony Nicole Perez (iD)

ABSTRACT

Understanding the impact of race on the enduring racial disparities and inequities throughout our institutions is a key tenet of social work competency. The purpose of this qualitative case study was to investigate the experiences of undergraduate social work educators (BSWEs) who teach the required diversity course. Participants reported the primary challenges teaching about race and racism were: (1) faculty racial identity and lack of credibility (2) emotional toll of teaching about race. This study has implications concerning educators' preparation to engage in anti-racist and equitable pedagogy in undergraduate social work education programs.

The murder of Trayvon Martin in 2012, COVID-19, and the 2020 summer of racial reckoning have proven to be historical markers of the resurgence of racial tensions in the United States. The recent killings of People of Color,[1] lack of legal repercussions, and subsequent local and national conflict have left little doubt that racism continues to be a pervasive problem in the United States. Racism is a system, rooted in an ideology of White Supremacy and reinforced by a continuous abuse of power to support prejudice and enforce discrimination. More than 50 years after the Civil Rights Act of 1964 passed, racial inequality and disparities persist in virtually every social aspect of the United States. Racial disparities in health, mortality and morbidity, behavioral health, academic and economic and criminal justice (Alexander, 2020; Fry et al., 2021; Gramlich, 2019) outcomes firmly mark racial injustices that exist in the U.S. This study was conducted during a time of intensified negative race relations in the United States with People of Color continuing to struggle on a daily for equity and equality and those who identify as White being barraged by a level of anger and frustration related to race they might find difficult to understand. The current socio-political climate exposed by social media illuminates the tenacious problem of race in the United States for the world to witness.

Social work is a practice-based profession and an academic discipline that promotes social change and development, social cohesion, and the empowerment and liberation of people. As such, the National Association of Social Workers (NASW) contend that social work educators should "promote sensitivity to and knowledge about oppression and cultural and ethnic diversity" (NASW, 2017, p. 5, Ethical Principles section, para. 3). However, undergraduate social work preparation programs have given little attention to advancing this goal. While social justice and cultural competency are foundational hallmarks of the social work profession, (National Association of Social Workers (NASW), 2021; Perez, 2021b), research shows very few educators have been provided any training to teach diversity courses (Holland, 2015; Perez, 2021a; Smith et al., 2017). Of the educators who have received training, it has been minimal at best (Perez, 2021b).

For social workers–and other helping professionals–experiences with racially diverse populations presents both opportunities and challenges for personal, professional, and organizational growth (Davis, 2016). By blending theory with a minimum of 400 mandatory field hours, undergraduate social work programs require students to put classroom knowledge into practice prior to graduation (Council on Social Work Education, 2015). Therefore, the expectation is at the baccalaureate-level students will have the basic knowledge, values, skills, and cognitive and affective processing to practice entry level social work upon graduation (CSWE, 2015).

By tending to crucial tasks, such as case management, biopsychosocial interviews, community outreach and advocacy, case worker, and juvenile court liaisons undergraduate social work degrees, are both useful and flexible (CSWE, 2015/2019; National Associations of Social Workers (NASW), 2015). Consequently, the opportunity for someone with a BSW degree to encounter the complexities of race and racism even prior to graduation in a professional setting is highly likely. Thus, the onus of the social work profession to strengthen their efforts to understand the function of race is central to addressing the call to promote sensitivity and knowledge around oppression and racial diversity. The purpose of this qualitative case study was to investigate the experiences of undergraduate social work educators who teach about race in the required diversity course. Therefore, the research question guiding this study was: what are the experiences of undergraduate social work (BSW) faculty who teach diversity courses?

Literature review

The Council on Social Work Education (CSWE), which guides social work education, mandate educators teach students to *"engage diversity and difference"* (CSWE, 2015, p. 7). Current literature directly focused on teaching about race and racism is limited within the field of social work education, particularly research on undergraduate social work education. Consequently, I also draw upon other bodies of knowledge including legal studies, teacher education, psychology, sociology, social justice education, and women's studies. Teaching race, racism, and anti-racism in higher education can present several challenges for educators. Researchers have documented student resistance to race-specific content may be the initial and most common barrier to engaging race, racism, and anti-racism that educators will encounter (Garner, 2017; Ladson-Billings & Tate, 1995; Perez, 2021b).

Student resistance and backlash

Jones (2008) defines student resistance as "resisting active engagement in issues related to diversity and social justice education and perceived challenges to prevailing conceptions of self, others, and compelling social issues of the day" (pp. 68–69). Resistance to race-specific content may manifest cognitively and/or emotionally as students grapple with racial content, such as power, privilege, and oppression, which are central ideas of diversity and cultural competence (Garcia & Van Soest, 1997; Perez, 2021a; Stewart, 2016; Tatum, 1992). Powerful emotional responses, which can range from guilt and shame to despair and anger, can lead to student resistance to the learning process (Johnston, 2011; Nadan & Stark, 2016; Smith et al., 2017; Tatum, 1992; Watt, 2007). Such resistance may interfere with students' cognitive processes as well as the mastery of the material.

Student resistance to discussions of race, racism, and anti-racism may take various behavioral forms in the classroom. Students may retreat and become silent (Author 2021, Henry, Cobb-Roberts, Dorn, Exum, Keller, and Shircliffe, 2007; Stewart, 2016; Tatum 1992), openly challenge the professor (Author, 2021; Henry et al, 2007), engage in micro aggressions (Henry et al, 2007; Matias, Henry, Darland, 2016; Sue, 2009), or any variety of negative behaviors. Similarly, student backlash can be a way to usurp the teacher-student power dynamics inherent in a classroom setting (Henry et al, 2007). Backlash may take the form of writing e-mails to the professor, expressing (or demanding) the student's point of view is correct and requesting a change in content, how it is presented, or discontinuation of the content/course altogether (Henry et al, 2007). If not met with the desired response from the professor, students may move their complaints up the chain of command and involve the dean or other administrators to assert their power (Henry et al, 2007; Matias et al., 2016). Additional challenges come by way of student backlash to the material presented. While resistance can be quieted or managed in the classroom, backlash spills over to other people in the department or the institution. How the administration handles student backlash demonstrates the amount of support faculty can count on. Lack of consistent application with the institution's legitimate grievance policy and lack of collaboration among departments could lead to negative learning experiences for students as well as faculty (Henry et al, 2007). However, another challenge that scholars have exposed is the lack of faculty pedagogical strategies and discomfort when teaching about race or racism.

Faculty lack of preparation and comfort as a barrier

Perez (2021b) asserts, "Anti-racism is not an affable topic yet if faculty approach the topics of race and racism with apprehension, the effect on the classroom can be detrimental to students' learning process around race" (p. 511). Smith et al. (2017) posits that educators are responsible for their continued development around issues of race. If faculty are not intentional about recognizing their process, they may find it more difficult to help students begin to embark on their journey. Faculty may experience feelings of discomfort including guilt, anger, frustration, and weariness with presenting race-specific content (Johnston, 2011; Smith et al., 2017). These emotions may manifest differently in Faculty of Color and White faculty. For the Faculty of Color, they could develop a sense of "cultural taxation" at being the go-to for teaching diversity or multiculturalism courses or being seen as the resident expert on all things race-related when a situation arises (Smith et al., 2017). White faculty, on the other hand, may experience a sense of defeat as they often lack support in developing an anti-racist identity, in both personal and professional realms (Holland, 2015; Johnston, 2011; Smith et al., 2017, Varghese, 2016).

All faculty, irrespective of race, face unique trials in teaching race, racism, and anti-racism. Faculty may struggle with questions of legitimacy due to their race or perceived race (Johnston, 2011; Smith et al., 2017). They must also deal with feelings of discomfort and their positionality in a classroom setting when teaching race, racism, and anti-racism (Johnston, 2011; Tuitt, Haynes, Stewart, 2018). As faculty contend with student resistance and backlash, lack of support from leadership, and learn how to navigate academia, they are often left with little or no training to deal specifically with teaching race, racism, and anti-racism (Perez, 2021a; Tuitt, Haynes, Stewart, 2018). To complicate matters more, they must err on the side of caution and remember that learning about race and racism is a journey not a place of arrival, which must be both intentional and life-long. Higher education is often the first opportunity students have to interact with diverse populations and expand their worldview. In remaining true to the commitment of social justice and engagement of difference, social work educators should reengage the impact of race within its programs.

Conceptual framework

I used a conceptual framework comprised of Critical Race Theory (CRT) and social justice pedagogy to design and analyze emergent data in this study. In her 2021 AERA Brown Lecture in Education Research, Patton Davis asserts that racism and white supremacy periodically mutates into variants that are more difficult to detect within society. CRT's basic premise that race is a defining characteristic in our institutions, laws, and policies as a result, racism is both inherent and normalized in U.S. society (Delgado, 1995; Ladson-Billings & Tate, 1995; Leonardo, 2004). Developed within legal studies, CRT has been increasingly applied to the field of education (Delgado & Stefancic, 2001; Ladson-Billings, 1998); however, it has been sparsely applied to the field of social work education. CRT provides an intersectional and in-depth analysis on the function of race and racism in the U.S.

In recent years, CRT has become a flashpoint in both political and social circles in the United States' continued reckoning with its racist past. Opponents of CRT have put forth that the theory seeks to rewrite American history, which is unacceptable, and rebukes *all* White people for being oppressors or that in turns condemns Black people to being hopeless victims. Detractors have focused crafting bans on teaching about race in the classroom, yet the fundamental issue with these claims are they remain at best an exaggeration of the theory. What is misunderstood is that CRT is traditionally not taught in K-12 schools (Ray and Gibbons, 2021, November; Patton Davis, 2021, October 21). The level of sophistication necessary to appropriately employ a CRT analysis requires advanced knowledge and skills that is typically developed within graduate education, not in the K12 years. Further, when applied, CRT can help foster the critical thinking skills to empower social workers to challenge and dismantle racists systems. CRT both names the systems and perpetrators of racism, allows space for the full and accurate teaching of history, and assists in moving the discourse beyond individuals. The focus of CRT is not at the individual level. At a structural level, CRT illuminates how racism is entrenched in U.S. society and enacted through policies and practices. It illuminates the dynamics of race (how race is understood, expressed, lived, etc.) on a macro-level, and how the current narrative of colorblindness obscures the power and privilege of the dominant group in the U.S. (Bonilla-Silva, 2018/2014; Delgado, 1995; Lopez, 2003). CRT seeks to dismantle race, gender, and class subordination by connecting theory to practice (Mensah Moore, 2016). Utilizing CRT, social work students may be able to recognize how race contributes to the persistent and increasing gaps in health and health care, wealth, and education.

Social justice pedagogy is a praxis of theory, reflection, and action where social justice principles are applied to all levels of education to expose unequal power positions (Adams, 2016, Perez, 2019). In this context, pedagogy denotes a holistic approach to the educational system, including curriculum and pedagogy, in need of critique and change (Adams, 2016; Apple et al., 2009). An aim of social justice pedagogy is to foster active engagement with social justice content to "affirm, model, sustain socially just learning environments for all participants" thus offering a liberatory model on how to engage in equitable relations in the larger society (Adams, 2016, p. 27). Adams (2016) argues that as social justice educators, the pedagogical choices made are as important as the content presented in the classroom. Moreover, Adams (2016) notes this anti-banking stance seeks to encourage students to become critical co-investigators in their learning process. It differs in traditional education by acknowledging the emotional component of learning new knowledge and the promotion of learning communities for students to learn from one another and apply that knowledge to larger systems (Adams, 2016).

Together, CRT and social justice pedagogy challenge social work educators to examine explicit and implicit curriculum as well as the cognitive and affective processes of students. Through this lens, I examined the challenges shared by the participants.

Methods

I recruited participants through professional contacts, at the 2018 Council on Social Work Education's Annual Program Meeting (APM), and social media outlets, which targeted social work educators. I also used purposive sampling (Lincoln & Guba, 1985; Patton, 2015; Shavers & Moore, 2014) to identify participants who met the following criteria: BSW faculty who (a) teach/have taught in the Southeast region of the U.S.; (b) teach/have taught a social work program's required diversity and/or social justice course; (c) had three or more years of experience teaching about race and racism; and (d) who had taught the required designated course a minimum of three terms. Nine (9) participants met the inclusion criteria. All participants chose their own pseudonyms (see, Table 1).

I conducted three semi-structured interviews with participants either face-to-face, video conferencing, or by phone (Glesne, 2015). Interviews took place from Fall 2018 – Spring 2019, lasted between an hour to 1.5 hours; length was largely dependent upon the respondents' willingness to share. Additionally, I collected course syllabi, power points, and assignment descriptions specific to race, racism, and anti-racism that the participants utilized in the classroom.

I then converted the recordings to audio recordings and transcribed them. To ensure accuracy, I compared each file to the original recording and validated it. The first coding cycle began with open coding technique (Saldaña, 2015) to get a sense of what the participant is sharing relative to teaching race and racism in the classroom. In the second coding cycle, I utilized focused coding to analyze how participants discussed strategies to respond to challenges of any kind (Saldaña, 2015). Finally, in the third level of coding I employed pattern matching using Dedoose to determine how findings align, or misalign, with the tenets of CRT, with competencies 2 and 3 presented by CSWE and the corresponding

Table 1. Participant demographics.

#	Degree	Pseudonym	Gender	Race	Licensure	Teaching	Teaching Diversity	PWI Public, Private or Religious	F/T Appt Type
1	PhD	Isabel	Woman	White	LCSW	5	4	Public	Adj.
2	PhD	Regal	Woman	B/AA	No	3	3	Public	TT
3	MSW	Wonder Woman	Woman	AA	LISW-S	4	3	Pub/Rel	Adj.
4	PhD	Cynthia	Woman	AA	No	25+	25+	Public	Tenure
5	MSW	Niang	Man	Senegalese	No	8	5	Public	NTT
6	PhD	Doc	Woman	AA	LMSW	18	3	Religious	TT
7	EdD	CJ	Woman	White	LSW	13	10	Private	Tenure
8	PhD	Taffy	Woman	White	No	19	3	Private	Tenure
9	PhD	Michael	Man	White	LISW-S	6	3	Public	TT

EPAS. I coded the data into themes utilizing the Council on Social Work Education EPAS. A pattern-coding procedure was utilized to bring structure and meaning to data and identify emergent themes and patterns (Patton, 2015; Saldaña, 2015). Informed by CRT, I also analyzed institutional arrangements that either contributed to or limited the delivery of race, racism, and anti-racism content with particular attention given to those who have less power in social and institutional settings (e.g., adjuncts, tenure-track faculty, and People of Color).

Furthermore, the syllabi, assignment descriptions, and PowerPoints began with coding focused by the EPAS composition of knowledge, values, and skills to systematically identify patterns and themes within the data. This was helpful as these categories put forth through the EPAS shape and drives course content as programs must meet these standards to earn accreditation (CSWE, 2008/2015). In the second phase of coding, I conducted a descriptive analysis and sort codes into the specific categories of knowledge, values, and skills and then looked for patterns within each category (Patton, 2015; Saldaña, 2015).

Themes

There were several challenges that emerged from the participants' interviews about their experiences when teaching about race and racism. Two significant challenges from these narratives were: (1) lack of credibility due to faculty racial identity and (2) emotional toll of teaching about race. The narratives of these participants illustrate how these challenges materialized within the social work classroom. Participants also shared the strategies they employed to overcome these challenges and facilitate positive cognitive and affective processing around the function of race in the United States. These selections represent only a portion of a more in-depth analysis of these participants' experiences with race and racism in the classroom.

Lack of credibility due to faculty racial identity

The racial identity of the faculty also served as an element that shaped the classroom environment. Both Black and White participants attributed their own racial identity as a barrier or adding to the challenge of teaching about race and racism. Taffy, a White female, also revealed, "One term I had a couple of students who sat in the back and just checked out. I once overheard them say I shouldn't even be teaching the class as a white lady." When asked, Taffy shared that the students were both "African-American and I just couldn't bridge that gap."

Niang, a Black Senegalese-American, disclosed, "I have been accused of being racist against whites because I challenge and disrupt their thinking around race. Students have gone to my dean saying I have an agenda, and she actually called me in to reprimand me. It is beyond frustrating, but I informed her that we have standards to stick to in social work and simply because that course is open to other majors doesn't change how I will teach it." Wonder Woman1920, a Black woman, experienced the questioning of her credibility in a less direct manner. She revealed, "I have come into class and had students question why I was writing on the board. Like they are surprised, I am the professor and are bold enough to challenge me. When we get to talking about race I usually have students question my motives but they do it in a roundabout way." When asked for clarification she shared, "Well they say things like, we've had a Black president, or identify the extreme racists as being fringe, or it's only a problem because people keep bringing it up."

For these participants their racial identity shaped the learning environment. Only one participant in the study identified benefits to his racial identity in the classroom environment. Michael shared, "I know I don't get questioned on things because I am a White male. It's not like my friend. She is an African American woman, and I don't get the same pushback. I really don't get any pushback." The others noted that in general they felt their racial identity had been a barrier they had to overcome each time they taught the course. For the White participants the lack of legitimacy was rooted in a lack of lived experience to be able to understand and communicate the experience of racism, while for the Black participants it was felt they had a hidden agenda and racism was not as impactful in modern society. These experiences often took an emotional drain on participants.

The emotional toll of teaching about race

The participants in the study unanimously agreed that to teach about race and racism is emotional and takes a toll. Michael insisted, "It is absolutely critical ... for social work students to wrestle with and come to a particular place, an anti-racist place, which is very challenging, emotionally, to do." Black faculty discussed the emotional toil teaching about race has on them. Regal, a Black woman, disclosed, "So one of the biggest barriers for me, it's really just having a talk about it and having to deal with my own emotions around a lot of these issues. My own emotions and talking about [race] is exhausting, really traumatizing to, you know, have to try to talk through issues related to race and deal with student resistance." Regal also shared, "There is no support. Nowhere to go to work through the daily trauma of trying to teach about racism and the push back." Likewise, Niang shared, "Students have gone to other faculty and the dean to complain. There is almost no support when a student complains. It's just what did you do wrong?" Regal and Niang both seem to infer that the emotional toll of teaching about race and racism was exasperated by the lack of institutional support.

Similarly, participants noted the increase in emotional trauma since the 2016 presidential election cycle. Cynthia shared, "It's exhausting and recently it has gotten scary," when asked what it was like to teach social work students about race. She expanded, "I teach in a small college town. When I taught the course this past term, some of the students got really bold and even disrespectful. I had to go to the dean for support, but nothing really happened. They let them stay in the class, and at the end of the term, my mailbox was vandalized. I thought seriously about retiring after that. I mean we know who it was by what they wrote, but there was no proof. And I just thought this job is not worth all this." Cynthia was visibly emotional when recounting this moment.

All the participants believed the current sociopolitical climate has contributed to the increasing difficulty in teaching students about race. Michael clearly illustrated this in his comment, "The current political climate, the current social climate is a barrier. Conversations about race and racism in the United States are emotional for good reason. There's outrage and injury from oppression, people are living it still today yet others are trying to deny it." When asked the impact of this, Michael shared, "It can be overwhelming. I mean after some classes I feel exhilarated because I see they got it and then others I feel frustrated that they aren't going deep." Another white faculty, CJ, admitted she is emotionally drained at times when teaching about race and racism. She stated, "As a white person I just want to get it right. I want other white people to understand how hurtful racism is and how it operates in the world. I see the harm; I have seen it in practice and see it on the news. And it is just getting worse since the [2016] election."

Niang further elucidated, "This socio-political climate makes it tough to teach about race. Students, particularly White students, try to negate or avoid it. As a Black man teaching about race in PWIs, students feel emboldened to challenge you. Actually, I hear that from other scholar educators of color in all college settings. There's just this boldness to defend their white fragility. It's tiring."

Black faculty more openly expressed the emotional toll teaching about race had on them. They also more often cited a lack of administrative and peer support that added significantly to the emotional fatigue they were experiencing. White participants most often cited the need for professional development and education in order to become better equipped at handling the emotional aspects around teaching about race and racism. All participants in this study shared that entering into the professoriate they were ill prepared to handle the emotional cost of teaching about race and racism in social work programs.

Discussion

This study illuminated the experiences of undergraduate social work educators (BSWE), which supported the need for an explicit insertion of critical analysis of race and racism within social work curriculum. Understanding how faculty operate in social work preparation programs allows for social work educators at various levels and institutional contexts to begin to shift the culture to include more critical, social justice oriented pedagogical strategies focused on race. Current social work education practice lacks a coherent and comprehensive approach to engaging racism or the necessary tools to support future educators in teaching such content. Thus, in effect preserving the very system the profession claims to work toward dismantling. Samaras et al. (2019) found that self-study can reinforce change over time, when examined from a CRT lens supports the idea that this method will only lead to small changes that will continue to fall short of the stated goal of developing social workers to be change agents against social injustice on a macro-level.

Another critical aspect discovered was the racial identity of the faculty and the connection to the levels of support they perceived, which aligns with the literature (Anft, 2018; Bathgate et al., 2019). Scholars note that teaching strategies are not only shaped by the faculty's personal beliefs and motivations but also by the departmental structure and the relationship with their colleagues (Brew & Mantai, 2017; Brownell and Tanner, 2015. The participants in this study described very different levels of support within their institutions. White faculty participants detailed ways in which their whiteness was enacted to provide a protective factor within their departments and their institutions.

Conversely, Black faculty with full-time status expressed concern about or acknowledged the consequences of teaching about race and racism. Noting the fact that student evaluations are a primary source used for tenure and promotion, the participants verbalized the potential negative impact student evaluations may have on their careers. The American Sociological Society (ASA) (2019) issued a formal statement calling student evaluations problematic, weakly related to student learning as well as biased against women and People of Color. The enactment of whiteness as property can be seen here as well. The White faculty did not mention the impact teaching about race could or did have on their career paths. Yet, even the Black faculty who had successfully reached tenured status comment on their negative teaching evaluations and the repercussions (reprimands by administrators, isolation by peers, student backlash) they experienced in their careers.

Study participants also shared the emotional currency they spent both inside and outside of the classroom due to teaching about race and racism. Inside of the classroom, most faced student resistance and white guilt while at times simultaneously trying to attend to the emotional needs of students both in the moment and after class. Furthermore, the lack of administrative support contributed to the emotional toll the faculty experienced when teaching about race and racism. For Black faculty participants this added a layer of pressure as well as stress when engaging race in the classroom that White participants did not convey. White participants similarly revealed an emotional toll of teaching about race and racism. However, their conflict had less to do with student resistance and was more of developing a strategy to supplement where they felt they were lacking experience with racism. Both Taffy and CJ expressed an internal conflict as they commented that teaching about race is challenging as a White person and feeling the strain to accurately portray the content, therefore, they shifted some of the responsibility to their Students of Color by asking them to provide examples of their lived experiences or of people, they may know in class. Yet, Black faculty spoke of intentionally protecting Students of Color from this very situation.

Social justice is a core value of the social work profession grounded in the history and development of the profession (NASW 2006, 2007, 2014, 2017). Bell (2016) notes that social justice is both a goal and a process. The *goal* is full and equitable participation of all people mutually shaped to meet their needs utilizing a democratic and participatory *process* for attaining the goal (Bell, 2016). Each of the participants in this study expressed a commitment to social justice as a process; how they educated students on the topics of race and racism, and as a goal; to develop students who engage in anti-racist practice.

One gap in meeting the CSWE mandate of teaching to engage diversity and difference is social work educators' potential lack of understanding of both pedagogical and andragogical strategies. Rahill et al. (2016) notes that social work students are becoming more diverse in age as well as with students returning to school later in life or to start a second career. Possessing knowledge regarding best practices in education can be pivotal when engaging students with difficult content such as race, racism, and anti-racism. This knowledge can assist social work programs in terms of course design, effective teaching strategies (including active learning), technology to help facilitate true learning, as well as challenges and solutions to presenting race-specific content. When educators have little to no training in pedagogical and andragogical strategies teaching race, racism, and anti-racism can be daunting. Varghese (2016) posited clinical social work educators often overlooked the impact of race on client symptomology in teaching students. For example, in overlooking race, social work educators may unintentionally perpetuate stereotypes such as the "strong Black woman" trope and misinterpret expressions of depression for resiliency. In his study, these educators often relied on or reverted to the prevailing Cultural Competency Model to assess clients, which unintentionally highlights stereotypical aspects of diverse populations, thus resulting in an inaccurate assessment of the etiology of symptoms as well as strategies to support them (Varghese, 2016).

Critical race theory (CRT) and social justice pedagogy in social work education

CRT provides a framework to study the relationship between race, racism, and power within social work education. Delgado and Stefanic (2012) described CRT as an activist movement grounded in legal scholarship that seeks to understand the effects of race and racism on society and change society for the better. For the participants in this study, the knowledge contributes both to how they conceptualize their effectiveness as an educator and assist how they support students' development. CRT challenges normative standards of whiteness and provides insight into how race, racism, and power interact to maintain and support racial inequality. CRT promotes a multi-directional, structural approach to addressing social injustice in a diverse society that goes beyond the expansion of access to opportunities.

Race and racism are decentralized and woven into the educational standards in such a way that celebrating difference and diversity is a substitute for critical engagement of the systemic impact racism has on the life conditions of Black and Brown people in the United States. Social work education's commitment to the multicultural approach highlights personal beliefs and attitudes as problematic and actively avoids racism as structure. This was a sentiment expressed by seven of the nine participants in this study. While this approach does challenge students to assess their personal values and views, it neglects the role racism places in creating racial disparities in the structure of society. Yet it allows social work education the appearance of critically attending to race and racism.

CRT offers a framework to go beyond the content integration, equity pedagogy, and prejudice reduction that may actually overemphasize differences and impede interrogation of the underlying systems and processes that create and maintain differences. Relative to the current study, the participants provided several examples of BSWEs expressing a desire to center race within social work education. Some participants specifically focused on students understanding race and racism from a historical standpoint and being able to analyze situations as well as apply their knowledge to their social work skillset. Moreover, they expressed being exposed to this knowledge in their educational programs would have provided the tools to support the development of their students as charged by CSWE.

The findings of this study also illuminated the importance of intentionality when teaching about race and racism. Participants shared their motivations for being intentional and their experiences in the learning environment. Specifically, the Black participants' motivations for being intentional were rooted in their own past educational experiences with race and racism in social work education. They reasoned that for social work students to progress in their understanding, self-awareness, and self-regulation about race, SWE need to present the importance of race early in students' educational careers. This mind-set highlights the importance of including voices of color, another CRT tenant, as a way to produce a more socially just program for social work students and their future clients.

Working in concert, social justice pedagogy prioritizes affirmation, modeling, and sustaining a socially just learning environment for all involved. This includes faculty situating themselves as mutual learners in the process. Faculty can utilize pedagogical strategies grounded in social justice principles. Some participants commented on the need to be mindful of where students often are developmentally when they enter the classroom. Isabel's comment conveys an acknowledgment of a holistic sense of learning about race and racism. The recognition and even use of emotion in the learning environment is an additional component of social justice education. As students encounter new perspectives and information, a myriad of emotions may arise. Social justice education actively seeks to engage emotion instead of shy away from it. There is a delicate balance at play to manage emotions that arise during discussions of race and racism without derailing the learning process. Encouraging students to look at different perspectives can be challenging, and the recognition of the emotional aspect of learning allows. These participants endeavor to create a learning environment for students and themselves to engage in meaningful exchanges to learn about the role of race in social work practice. To do so they employ several social justice education pedagogical strategies and principles.

Recommendations and future research

Based on the discoveries of this study, there are several recommendations and opportunities for future research. This will allow for a more in-depth and well-rounded perspective of race and racism in undergraduate social work education.

(1) Employ a comprehensive approach that incorporates both a course focused on race and racism, and anti-racist content scaffolded throughout the curriculum.
(2) Undergraduate social work programs should develop a required course on race and racism offered early in a student's academic track. This course could provide context to the history of race and racism from a social work lens specifically incorporating social work knowledge and values.
(3) Due to the unique skills required to effectively facilitate learning around race toward a social justice model, additional opportunities should be presented through CSWE, BPD, NASW, and other organizations to provide training and professional development to both full-time and contingent faculty.

Additionally, future research could go in several potential directions.

(1) Research focused on the development of social work pedagogical strategies explicitly grounded in social justice, race based, and critical foundations.
(2) Research designed to explore the experiences of field educators with race and racism during internship placements.
(3) Conduct a study focused on undergraduate social work students' experiences with anti-racist content in their undergraduate programs.

Conclusion

Social workers are educated to be practitioners to work in communities as boots on the ground professionals to help alleviate America's social problems. The lack of education about race and racism is an oddity considering social workers address , practically, every social problem and practice in diverse communities. Simply ignoring or rendering race and racism to be a problem for only People of Color instead of an issue for Whites as well holds the racial hierarchy in place. Social work preparation programs need to reimagine revolutionary ways to challenge systems that impede the development of anti-racist social work praxis.

Note

1. Capitalization of terms is not neutral and can denote solidarity in representation. The term "People of Color" highlights the solidarity of Asian, Black, Indigenous, and Latinx, and Multiracial people. Given this, I have chosen to capitalize collective nouns that reference specific groups which have been collectively marginalized and seek sociopolitical unity, and power. In all other instances, the terms utilized are those the authors applied in their work.

Disclosure statement

No potential conflict of interest was reported by the author(s).

ORCID

Ebony Nicole Perez (ID) http://orcid.org/0000-0002-8887-1278

References

Adams, M. (2016). Pedagogical foundations for social justice education. In M. Adams & L. A. Bell (Eds.), *Teaching for diversity and social justice* (pp. 27–54). Routledge.

Alexander, M. (2020). The new Jim Crow: Mass incarceration in the age of colorblindness- 10th anniversary edition.

American Sociological Society. (2019). Statement on Student Evaluations of Teaching. https://www.asanet.org/sites/default/files/asa_statement_on_student_evaluations_of_teaching_feb132020.pdf

Anft, M. (2018, December 21). *Colleges step up professional development for adjuncts.* Chronicle of Higher Education.

Apple, M. W., Au, W., & Gandin, L. A . (2009). Mapping critical education. In Michael W. Apple, Wayne Au, and Luis Armando Gandin, eds., *The Routledge International Handbook of Critical Education*, (pp. 3–19). New York: Routledge.

Bathgate, M. E., Aragón, O. R., Cavanagh, A. J., Frederick, J., & Graham, M. J. (2019). Supports: A key factor in faculty implementation of evidence-based teaching. *CBE - Life Sciences Education, 18*(2), ar22. https://doi.org/10.1187/cbe. 17-12-0272

Bell, L.A. (2016). Theoretical foundations for social justice education. In M. Adams Editor & L. A. Bell Editor, *Teaching for diversity and social justice* (pp. 3–26).

Bonilla-Silva, E. (2014). *Racism without racists: Color-blind racism and the persistence of racial inequality in America.* Rowman & Littlefield.

Bonilla-Silva, E. (2018). *Racism without racists: Color-blind racism and the persistence of racial inequality in America.* Rowman & Littlefield.

Brew, A., & Mantai, L. (2017). Academics' perceptions of the challenges and barriers to implementing research-based experiences for undergraduates. *Teaching in Higher Education, 22*(5), 551–568. https://doi.org/10.1080/13562517. 2016.1273216

Council on Social Work Education. (2008). Educational Policy and Accreditation Standards.

Council on Social Work Education. (2015). Educational policy and accreditation standards.

Council on Social Work Education. (2019). *2018 annual statistics on social work education in the United States.*

Davis, L. E. (2016). Race: America's grand challenge. *Journal of the Society for Social Work and Research, 7*(2), 395–403. https://doi.org/10.1086/686296

Delgado, R. (Ed.). (1995). *Critical race theory: The cutting edge.* Temple University Press.

Fry, R., Bennett, J., & Barroso, A. (2021). *Racial and ethnic gaps in the U.S. persist on key demographic indicators.* Pew Research Center.

Garcia, B., & Van Soest, D. (1997). Changing perceptions of diversity and oppression: MSW students discuss the effects of a required course. Journal of Social Work Education, 33(1), 119–129.

Garner, S. (2017). Racisms: an introduction. Los Angeles: Sage.

Glesne, C. (2015). *Becoming qualitative researchers: An introduction* (5th ed.). Pearson.

Gramlich, J. (2019). *From police to parole, black and white Americans differ widely in their views of criminal justice system.* Pew Research Center.

Hannah-Jones, N. (2014, May). Segregation now. The Atlantic. http://www.theatlantic.com/magazine/archive/2014/05/segregation-now/359813/

Henry, W. J., Cobb-Roberts, D., Dorn, S., Exum, H. A., Keller, H., & Shircliffe, B. (2007). When the dialogue becomes too difficult: A case study of resistance and backlash. *College student affairs journal*, 26(2), 160.

Holland, A. E. (2015). The lived experience of teaching about race in cultural nursing education. *Journal of Transcultural Nursing*, 26(1), 92–100.

Johnston, A. (2011). White professors, students of color, teaching race. Retrieved from http://studentactivism.net/2011/01/06teaching-race-1 & 2

Jones, S. R. (2008). Student resistance to cross-cultural engagement: Annoying distraction or site for transformative learning? In S. R. Harper (Ed.), Creating inclusive campus environments for cross-cultural learning and student engagement (pp. 67–85). Washington, DC: NASPA.

Ladson-Billings, G. (1998). Just what is critical race theory and what's it doing in a nice field like education? *International Journal of Qualitative Studies in Education*, 11(1), 7–24.

Ladson-Billings, G., & Tate, W. F. (1995). Toward a critical race theory of education. *Teachers College Record*, 97(1), 47–68. https://doi.org/10.1177/016146819509700104

Leonardo, Z. (2004). The color of supremacy: Beyond the discourse of "white privilege. *Educational Philosophy and Theory*, 36(2), 138–152. https://doi.org/10.1111/j.1469-5812.2004.00057.x

Lincoln, Y. S., & Guba, E. G. (1985). *Naturalistic inquiry.* Sage Publications.

Lopez, G. (2003). The (racially neutral) politics of education: A critical race theory perspective. *Educational Administration Quarterly*, 39(1), 68–94. https://doi.org/10.1177/0013161X02239761

Matias, C.E., Henry, A., & Darland, C. (2016). The twin tales of whiteness exploring the emotional roller coaster of teaching and learning about whiteness. *Taboo: The Journal of Culture and Education.* 16(1). Retrieved from https://digitalcommons.lsu.edu/ taboo/vol16/iss1/4

Mensah Moore, F. (2016). Preparing for discussions on race and racism: The critical voices in teacher education course. [Paper Presentation] *NARST International Annual Conference*, Baltimore, MD, United States.

Nadan, Y., & Stark, M. (2016). The Pedagogy of Discomfort: Enhancing Reflectivity on Stereotypes and Bias. *British Journal of Social Work*, 47(3), 683–700.

National Association of Social Workers (NASW). (2021). *Code of ethics.* NASW Press.

National Associations of Social Workers (NASW). (2015). *Standards and indicators for cultural competence in social work practice.* NASW Press.

National Associations of Social Workers (NASW) (2017) NASW code of ethics. Washington, DC: NASW Press.

Patton, M. Q. (2015). *Qualitative research & evaluation methods: Integrating theory and practice.* SAGE Publications, Inc.

Patton Davis, L. (2021, October 21). Still Climbing the Hill: Intersectional Reflections on Brwon and Beyond. [AERA : 18th Annual Brown Lecture in Education Research]. *Virtual.* https://youtu.be/EE25GVC2_1Y

Perez, E.N. (2019). Using Critical Race Theory to Examine Race and Racism in Social Work Education. [Unpublished doctoral dissertation]. University of South Florida.

Perez, E. N. (2021a). Teaching race and racism in social work education. In L. Parson & C. C. Ozaki (Eds.), *Teaching and learning for social justice and equity in education* (pp 177–197). Palgrave McMillian.

Perez, E. N. (2021b). Faculty as a barrier to dismantling racism in social work education. *Advances in Social Work*, 21(2/3), 500–521. https://doi.org/10.18060/24178

Rahill, G. J., Joshi, M., Lucio, R., Bristol, B., Dionne, A., & Hamilton, A. (2016). Assessing the development of cultural proficiency among upper-level social work students. *Journal of Social Work Education*, 52(2), 198–213.

Ray, R. and Gibbons. A. (2021, November). Why are states banning critical race theory? *Brookings.* https://www.brookings.edu/blog/fixgov/2021/07/02/why-are-states-banning-critical-race-theory/

rownell, S. E., & Tanner, K. D. (2015). Barriers to faculty pedagogical change: Lack of training.

Saldaña, J. (2015). *The coding manual for qualitative researchers.* Sage.

Samaras, A. P., Hjalmarson, M., Bland, L. C., Nelson, J. K., & Christopher, E. K. (2019). Self-study as a method for engaging STEM faculty in transformative change to improve teaching. *International Journal of Teaching & Learning in Higher Education*, 3(2), 195–213.

Shavers, M. C., & Moore, J. L. (2014). Black female voices: Self-presentation strategies in doctoral programs at predominately white institutions. *Journal of College Student Development*, 55(4), 391–407. https://doi.org/10.1353/csd.2014.0040

Smith, L., Kashubeck-West, S., Payton, G., & Adams, E. (2017). White professors teaching about racism: Challenges and rewards. *The Counseling Psychologist*, 45(5), 651–668. https://doi.org/10.1177/0011000017717705

Tatum, B. (1992). Talking about race, learning about racism: The application of racial identity development theory in the classroom. *Harvard Educational Review, 62*(1), 1–25.

Tuitt, F., Haynes, C., & Stewart, S. (2018). Transforming the classroom at traditionally white institutions to make black lives matter. *To Improve the Academy, 37*(1), 63–76. https://doi.org/10.1002/tia2.20071

Varghese, R. (2016). Teaching to Transform? Addressing Race and Racism in the Teaching of Clinical Social Work Practice. *Journal of Social Work Education, 52*(sup1), S134–S147.

Watt, S. K. (2007). Difficult dialogues, privilege and social justice: Uses of the Privileged Identity Exploration (PIE) model in student affairs practice. *The College Student Affairs Journal, 26*, 114–126.

Part II

Conceptualizing anti-racist social work practice and research

Dual pandemics or a syndemic? Racism, COVID-19, and opportunities for antiracist social work

Kimberly D. Hudson (ID), Sameena Azhar (ID), Rahbel Rahman (ID), Elizabeth B. Matthews and Abigail M. Ross (ID)

ABSTRACT

In this article, we critically engage the "dual pandemics" framing of this special issue. We first consider the key assumptions of this popular frame, specifically the conceptualization of racism as a pandemic, and examine limitations of medicalizing racism. We follow with an introduction of the term *syndemic*, coined by public health scholar Merrill Singer, and discuss how the language of syndemics might accurately characterize the synergism and interconnectedness of racism and COVID-19. We conclude by applying syndemic theory to offer insights and opportunities for social work research, practice, and policy from a racial justice lens.

At the time of writing, the Centers for Disease Control and Prevention (CDC) estimate that over 834,077 lives have been lost to COVID-19 in the United States (U.S.; CDC, 2021a). In light of the profound health, mental health, and economic impacts of this pandemic, major public health, epidemiological, and medical societies have declared the COVID-19 pandemic a national emergency (Krieger, 2020). Racial disparities in COVID-19-related morbidity and mortality have been widely recognized and reported; individuals from minoritized racial and ethnic groups have disproportionately higher rates of COVID-19 infection and are more likely to die as a result of COVID-19 than white Americans (Anyane Yeboa et al., 2020). Similar patterns of disparate health care access and outcomes have been well documented the U.S., and have been maintained and exacerbated by the country's long history of injustice, structural racism, and interpersonal bias (Krishnan et al., 2020). This inequity is not confined to health or health systems; manifestations of structural, institutional, and individual racism have resulted in unequal access to employment and resources, failing public school systems, unsafe neighborhoods, food deserts, targeted mass incarceration, police brutality, maternal and infant mortality, obesity, chronic health conditions, among others (Johnson-Agbakwu et al., 2022; Thakur et al., 2020).

In recognition of this widespread phenomenon, leaders of 8 states, 105 cities and 76 counties in the U.S. have declared racism a public health crisis (American Public Health Association [APHA], 2021). This discourse positions racism and COVID-19 as two emergencies that are simultaneously afflicting the country. In an effort to understand and describe these intersecting crises, experts in policy and practice have argued that the United States is facing *dual pandemics*, a conceptualization of the novel coronavirus and structural racism as two deadly and widespread diseases (Martínez et al., 2021; Tan et al., 2021).

The concept of dual pandemics has been invoked across academic disciplines, such as medicine, public health, and social work (including in the title of this special issue), as well as in the public sphere in several ways. First, the concept of dual pandemics is used to explain the unique, present moment characterized by a mainstream (white, U.S.-born) awakening to the nefarious effects of racism. From this perspective, racial disparities in COVID-19 morbidity and mortality rates coupled with the murders of George Floyd, Breonna Taylor, Ahmaud Arbery and other Black Americans during the first half of 2020 created a "tipping point" (Brodie et al., 2021) that has led to unprecedented levels of mobilization around issues of racial equity. Some polls estimate that over 23 million Americans participated in racial justice protests and rallies, events that were largely organized by grassroots social movements such as Black Lives Matter, making this the largest protest movement in U.S. history (Buchanan et al., 2020). To this end, the concept of dual pandemics describes current conversations around racism and race relations by framing them as a product of the unique convergence of the COVID-19 pandemic and a series of highly publicized instances of police violence. A second purpose behind the use of the dual pandemics terminology is to describe the ways in which COVID-19 has laid bare the deeply entrenched structural racism that pervades all U.S.-based social systems. In contrast with the dual pandemics conceptualization that serves to encapsulate a novel moment in time, this second use of the concept of dual pandemics highlights existing and persistent problems, arguing that inequities in both COVID-19 morbidity/mortality rates and COVID-19 vaccine access are instead symptomatic of the long-standing problems of structural racism that have plagued the U.S. for centuries.

In this article, we seek to explore the benefits and drawbacks of conceptualizing COVID-19 and racism as dual pandemics by first considering the disease model of racism that is inherent in this language, and subsequently examining the dangers of medicalizing racism. We then introduce Merrill Singer's (2009) concept of *syndemic* in order to situate the dynamic, interwoven relationships among racism, disease, and their contexts. We draw on social determinants of health (SDOH) and political economy of health (PEH) models to illustrate how syndemic theory can be applied to understand disparities in vaccinations as rooted in medical mistrust and mistreatment. We close with a reflection on how a syndemic frame may more appropriately inform racial justice solutions in policy and social work practice.

The disease model of racism and its limitations

Characterizing racism as a pandemic requires one to envision racism in medicalized terms – to see racism as a disease. The idea of racism as a disease has roots tracing back to the post-World War II era. Bolstered by the increasing authority of the disciplines of psychology and psychiatry, this period was marked by a growing interest in identifying both medical and sociological explanations for human behavior (Thomas, 2014). Under this medical lens, the inhumane treatment of Jewish populations during World War II began to be viewed as psychologically abnormal; similar conceptualizations were later applied within the United States to understand the country's long-standing discrimination against African American populations, gradually evolving into a theory that described racism as a psychological illness (Thomas & Byrd, 2016).

Racism in the United States has been described using various medicalized terminology, including a mental or behavioral disorder, an infection, a social cancer, and as an addiction that requires individual and institutional recovery (Dobbins & Skillings, 2000; Kahn, 2017; Thomas, 2014). This disease model of racism draws parallels with established psychiatric diagnostic criteria found in the American Psychiatric Association's (APA) Diagnostic and Statistical Manual (DSM), asserting that racism is often "unwittingly and unwillingly" (Skillings & Dobbins, 1991, p. 206) hosted by afflicted individuals and causing persistent distress or disability to individuals, to communities of color, and to U.S. society more broadly (Poussaint, 1999). Some debates have considered the inclusion of bigotry and racism in the DSM, though action in this regard has not yet been taken (Bell & Dunbar, 2012).

While a medicalized approach may make academic conversations about racism more accessible, and also has been an appealing paradigm in light of the convergence of COVID-19 and racialized violence and protests, this approach has been critiqued for its shortcomings and dangers. As Kahn (2017) suggests, the medicalization of racism carries multiple risks, including removing the issue of racism from the public realm and into the professional space; positioning racism as an individual-level issue; erasing its structural roots; and depoliticizing racism by absolving individuals of their responsibility to personally address it. Ultimately, the disease model provides a limited and insufficient framework for addressing the intersection of race, racism, and COVID-19. Notions of "illness" and "disease" suggest that racism exists as an exogenous threat that has somehow invaded or afflicted an otherwise "non-racist" self. Individuals perpetuating racist ideas, beliefs, or structures therefore maintain the ability to decouple these behaviors or cognitions from those defining their core identity, and instead situate racism as a physiological or neurological abnormality that can be examined as a discrete – and potentially curable – phenomenon (Kahn, 2017).

In particular, the disease model suggests that racism is best "cured" through psychological, psychiatric, or neurological intervention at the individual level (Kahn, 2017; Thomas, 2014). By limiting the problem of racism to a function of an individual's cognitive processes (e.g., maladaptive attitudes and socially incongruent beliefs), this paradigm deemphasizes the role of larger political and social structures that have been built in service of upholding oppressive practices and will continue to perpetuate inequities, regardless of implicit or explicit racist cognitions (Wellman, 2000). Indeed, rather than an abnormal psychological state, many have argued that racism is so deeply entrenched into the fabric of our society that it is the rule, rather than the exception (Poussaint, 1999). For this reason, though corrective action at the individual level is a necessary component of antiracist efforts, it is an insufficient strategy to dismantle a system of thinking and a social structure that has been shaped by centuries of oppressive social policies and practices. Instead of viewing antiracist work as a collective, social responsibility of the public, the disease model suggests that addressing racism is best achieved by identifying and curing "infected" individuals (Wellman, 2000) and by labeling the "symptoms" of individual-level prejudice. It is through this lens that the recent murders of Black Americans at the hands of police are framed as a series of unique and troubling instances of novel racism committed by a few "bad apples." Similarly, this thinking enables racial disparities in COVID-19 infection and mortality to be attributed to discrete, poorly implemented policy or messaging choices by misinformed actors within the White House and the CDC, rather than a reflection of deeply and historically entrenched patterns of violence and oppression.

Conceptualizing COVID-19 and racism using a syndemic frame

An alternative conceptualization to the dual pandemic paradigm is offered through the lens of syndemic theory. Originally coined by Merrill Singer (2009) to describe concurrent issues impacting the HIV epidemic, the term syndemic refers to the phenomenon wherein health problems interact synergistically to create health disparities among particular communities. Syndemics are characterized by interactions between diseases and the contexts within which they are experienced that increase susceptibility for poor health outcomes (Horton, 2020). The notion of a syndemic focuses on explaining why diseases cluster, recognizing the political and social factors that transpire across generations that drive this clustering and create lasting power inequities due to systemic failures (Gravlee, 2020; Mendenhall, 2020).

Another key distinction between dual pandemics and a syndemic is that the latter does not construct racism as a virus. Under this view, racism cannot be isolated in a Petri dish like a microbe; it is pervasive throughout American social institutions, structures and systems. Structural racism is not a foreign body obstruction; it is a homegrown phenomenon carefully implemented and monitored over centuries, that serves the interest of the ruling class and specifically white supremacy and patriarchy. There is no mask that will prevent the damaging effects of racism; no antiretroviral medication to treat this "infection;" no vaccine to prevent its recurrence. By moving away from a dual pandemic concept, contemporary discourse around racism and COVID-19 can avoid the pitfalls of medicalizing racism, which may encourage an individualistic framing of complex, intersectional social issues, or suggest the availability of a simple and easy "cure" to this condition, such that attending a diversity and inclusion training would provide immunity from reproducing racism in our own personal and professional interactions, either consciously or unconsciously. Moving away from this individualized, medicalized framing may be necessary to encourage efforts to redress inequities that seek to reconfigure existing social structures and the requisite redistribution of power and resources.

An additional benefit of the concept of a syndemic is that it may better explain the mutual risk shared by seemingly disparate populations. Exposure to disease is operationalized through engagement in certain "risky" behaviors, while prevention is operationalized through risk-minimizing, behavioral strategies. For HIV, those risky behaviors were unprotected sex and shared needle use; for COVID-19 they are being unmasked, sitting indoors with people outside of one's household for long periods of time, attending large public gatherings, or not being vaccinated. Yet epidemics, including HIV and COVID-19, disproportionately impact those already socially marginalized for other reasons, including racism, poverty, immigration status and citizenship. Individual-level behaviors are therefore insufficient in explaining the disproportionate burden of disease that falls on particular populations locally and globally. For example, even within states where risk mitigation strategies (e.g., mask mandates or travel restrictions) were universally and aggressively enforced, COVID-19 continued to disproportionately affect individuals occupying certain social positions, including incarcerated people in jails or prisons, nursing home residents, and occupants of crowded housing in low-income neighborhoods. To understand and explain these patterns, the syndemic frame invites an examination of what social, political, or other forces shape the unique intersection of these social positions and COVID-19 risk, viewing these things as necessarily interconnected, rather than parallel to one another.

A syndemic framing of COVID-19 and racism advances racial justice in three ways. First, it shifts the culpability of the spread of disease and racism alike onto the state and other structural factors, rather than situating this as an affliction of an individual. Second, the syndemic frame invites a more nuanced understanding of how historical sociopolitical events have necessarily shaped the particular impact of COVID-19 on communities of color, and how we currently risk replicating harmful and patterns that have exacerbated disparate health outcomes and oppressive social systems by adopting a dual pandemic lens. Third, clarifying linkages between syndemic theory and the on-the-ground patterns of racism and pandemics, past and present, illuminates specific, actionable interventions toward antiracist futures.

Social determinants, political economies, and epidemics past and present

Syndemic theory requires one to consider not only the interactions between disease elements, but also the social, political, and economic contexts in ways a dual pandemic frame may fail to address, particularly if racism is not explicitly characterized as structural. Syndemic theory allows for insight into racial disparities in COVID-19 incidence by demanding attention to social determinants of health (SDOH) and political economies of health (PEH). The SDOH model serves the syndemic frame by describing how one's environment shapes health outcomes, while the PEH model explains how deeply rooted systems of power and privilege shape how policies and practices are implemented, and whose interest they serve.

Social determinants of health

SDOH, defined as the conditions in which we are born, live, work and play (Office of Disease Prevention and Health Promotion, 2018), help to explain how particular communities of color have been disproportionately affected by COVID-19 and other past epidemics, such as the Spanish Flu and H1N1. SDOH include access and quality of education, access and quality of health care, neighborhood and built environment, social and community contexts, and economic stability. Shared risk factors rooted in the SDOH impact a number of chronic health issues and have been co-occurring across epidemics (Marmot & Wilkinson, 2005).

SDOH have impacted health outcomes for current and past disease outbreaks alike (Abrams & Szefler, 2020; DeBruin et al., 2012). Specifically, SDOH inequities in the context of COVID-19, H1N1 and the Spanish Flu have consistently manifested through poor access to health care, lack of regular primary care providers, biased attitudes of health care workers, low health care literacy, reliance on public transportation, employment in positions that increase exposure to the virus, and ongoing residential segregation that causes communities of color to live in densely populated areas (McCoy, 2020; Thakur et al., 2020). Additionally, individuals belonging to racial/ethnic minoritized groups are more likely than their white counterparts to have occupations within essential services or be frontline healthcare workers (Devakumar et al., 2020). Similar to the case with H1N1, many could not work from home during COVID-19 surges. Fear of job loss and/or dependence on an hourly wage income left many of these workers vulnerable to working through the COVID-19 surges with limited personal protective equipment (PPE; Thakur et al., 2020). These workers also tend to have low-paying jobs and limited health care insurance, resulting in economic destabilization when succumbing to the illness. These working conditions are in turn associated with a higher likelihood for underlying health conditions (e.g., hypertension, diabetes, and obesity) that increase risk for COVID-19 morbidity and mortality (Abrams & Szefler, 2020; Laurencin & McClinton, 2020).

A similar detrimental economic impact was felt by Black people during the Spanish Flu epidemic (McCoy, 2020). Using a syndemic frame, a historical reckoning with the events of the Spanish Flu can illustrate how such social and political forces shaped the course of the pandemic, specifically by reliance on underestimated race-stratified data, missing records, misdiagnosis, and underreporting (Krishnan et al., 2020). Notably, during the Spanish Flu epidemic of 1918, several majority-Black states in the South were not covered in the National Births and Deaths Registration Area (Økland & Mamelund, 2019) and were not included in the mortality records associated with the Spanish Flu. Due to overt racism and segregation, cases of African Americans were not treated by physicians due to poverty, discrimination by white doctors, lack of access to Black doctors, or poor access to health care, health information, and health insurance (Økland & Mamelund, 2019). Black communities were left to take care of themselves, often providing medical care in basements and hospitals with inadequate support (McCoy, 2020), leading to many preventable deaths. The collective precarity that continues to be experienced by communities of color across multiple contagious diseases is testament to the impact of the social forces of racism, which predates any individual outbreak or epidemic.

Political economies of health

Similar to the SDOH model, the PEH model considers how political and economic domains interact to shape individual and population health (Harvey, 2021). In the current paradigm, the key economic and political tenets of American neoliberalism made it possible for an infectious disease pandemic to spread, unchecked, in the United States. According to the World Bank (2018), the U.S. spends an average of $10,624 per year per capita on health care, grossly outspending other Western industrialized countries, such as Germany ($5,472), Canada ($4,995) and the United Kingdom (4,315). Nonetheless, the United States has held one of the greatest global burdens for COVID-19 (Centers for Disease Control and Prevention [CDC], 2021b). This disparity may be the logical consequence of a PEH that has largely prioritized and rewarded financial outcomes over health outcomes, or alternatively prioritized and rewarded maximization of profit over the max-imization of health. Within the privatized managed care system in the U.S., for example, cost containment, and differential reimbursement rates by public and private payers have led to the closures of hospitals that primarily serve communities of color (Bailey & Moon, 2020) and the regulation of communities of color to certain healthcare settings. In the past few years there has also been unprecedented focus and mobilization of resources in addressing white "deaths of despair" (Bailey & Moon, 2020). This phenomenon defines racial capitalism whereby white lives are impli-citly valued over others; and the suffering and death of Black, Native American, and Latinx people are normalized (Bailey & Moon, 2020).

A closer look at the COVID-19 pandemic: medical mistrust, mistreatment, and vaccine hesitancy

A syndemic frame centers the SDOH and PEH, helping explain current racial injustice related to the COVID-19 pandemic. Disparate COVID-19 vaccination rates, for example, are best understood not as individual choices or preferences against receiving the vaccine, but as shaped by environmental, social, and economic barriers, as well as deeply entrenched medical mistrust caused by generations of mistreatment from the medical community. As efforts to provide COVID-19 testing, treatment and vaccination have ensued over the past year, some communities of color have been hesitant to trust public health measures (Bogart et al., 2021). In recent months, Black and Latinx people have been less likely to be vaccinated compared to their shares of COVID-19 cases and deaths (Kaiser Family Foundation [KFF], 2021). Since March 1, 2021, the proportion of the population who has received at least one dose of the COVID-19 vaccine has increased across the general population, but disparities for Black and Latinx people have widened (KFF, 2021).

Syndemic theory is a useful mechanism to understand how contemporary and historical social and economic determinants, grounded in racist systems and structures, contribute to this disparity. These trends may be linked to medical mistrust of the experimental medications used in vaccines and to treat advanced COVID-19 respiratory disease (Bogart et al., 2021). Medical mistrust among Black, Indigenous and Latinx adults could also be due to the limited information on viral transmission and preventive measures received during the early stages of the pandemic, further posing risk for COVID-19 transmission (Leitch et al., 2020). This is despite the fact that misinformation and erroneous narratives about Black immunity have circulated through social media (Krishnan et al., 2020). However, it is important to note the distinction between scientifically proving vaccine efficacy and demonstrating community trustworthiness (Bunch, 2021). While vaccine efficacy can be demonstrated through reference to biostatistics and epidemiological studies, medical mistrust stems from historically rooted skepticism regarding an institution's motives (Bunch, 2021). Therefore, simply educating minoritized groups about vaccine efficacy will be insufficient in changing institutional mistrust.

Others have argued that challenges regarding vaccine uptake are generally less about vaccine hesitancy and more about barriers to access (Bedford et al., 2018). While vaccine hesitancy frames the problem of the lack of widespread vaccination to be the fault of individual actors, vaccine justice speaks to larger issues of equity regarding community trust, access and distribution. As per CDC guidelines (2021c), COVID-19 vaccines are to be provided at 100% no cost to all people living in the United States, regardless of their immigration or health insurance status. Yet underlying structural inequities have created increased barriers to vaccine access for people of color and people living in rural areas. Trends of reduced vaccine uptake among communities of color may also be the historical legacy of centuries of medical mistreatment and malpractice.

Even prior to the COVID-19 pandemic, Black people in the U.S. have been more likely than white people to have had their pain denied, their medical conditions misdiagnosed, and necessary treatments withheld by physicians (Bajaj & Stanford, 2021). Conversely, in a study of racial biases among clinicians, white patients were viewed to be significantly more likely to improve, to adhere to treatment, and to be personally responsible for their health (Khosla et al., 2018). These biases affect the ways in which people of color are treated, and in turn expect to be treated, within medical and public health institutions that purport to care for their wellbeing. Legacies of institutionalized racism may also trace back to ethical violations in research, ranging from the usage of Henrietta Lacks' genetic information without her consent to the denial of medical treatment to participants in the Tuskegee Syphilis Study (Azhar & DeLoach McCutcheon, 2021; Bajaj & Stanford, 2021). Witnessing and re-witnessing acts of police brutality on the news and social media can also cause vicarious trauma for communities of color, which may also impact levels of trust in medical and public health institutions (Laurencin & Walker, 2020).

Implications and opportunities of a syndemic frame

Using syndemic theory to frame the interconnected and mutually potentiating relationships among COVID-19 and racism, attending to both disease and environment elements, draws attention to the manifestation of these issues across individual, interpersonal, social, structural, and ideological levels. This multilevel reconceptualization invites opportunities for more complex racial justice strategies for exploring and examining concomitant, intersecting issues, developing multidimensional solutions, and delivering strategic interventions. To paraphrase Dr. Kimberlé Crenshaw, a legal scholar who coined intersectionality theory, "how you see a problem is how you solve a problem" (Crenshaw, 2016). Syndemic theory, as a general framework, can guide a number of interventions and serve as a bridge to antiracist futures. The following section will explore specific insights and specific pathways for action. As a discipline that encompasses many areas of practice, enacting meaningful change within the social work profession requires attention to the many contexts in which social workers learn, work, and advocate. In an acknowledgment of the breadth of our field, we offer recommendations uniquely tailored to domains of social work scholarship, education, practice and policy. Discrete recommendations and examples of associated action steps are summarized in Table 1 to illustrate how these changes could be carried forward within the discipline.

Table 1. Opportunity domains, recommendations, and specific examples of action steps.

Domain	Recommendations	Specific examples
Scholarship	Expand set of conceptual tools and integrate multiple levels of analysis	Deepen attention to intersectionality across all aspects of research (conceptualization, articulation of results, implications and dissemination)
		Contextualize findings in structural and social factors rather than characteristics of minoritized groups.
		Examine chronicity and the experience of ill-health within minoritized groups
		Allocate resources to examine race-stratified data
	Build awareness of how positionality shapes one's assumptions and worldviews	Incorporate researcher reflexivity
	Challenge traditional forms of positivist research paradigms in favor of those that engage critical consciousness-raising	Use Community Based Participatory Research to inform research and program priorities
		Revising aims and scope of academic journals to include submissions of other methodologies besides empirical research
		Encourage editorial boards to intervene on practices that result in racial inequity of scholarship
Education	Revise social work "canon"	Teach about the racist legacies of the profession
		Include narratives that reflect experiences of minoritized educators and practitioners
		Develop and evaluate courses, assignments, and field placements focused on antiracist social work practice
	Create a pipeline of future social workers and social work scholars of color	Hire faculty who represent minoritized communities
		Increase resources in mentorship and support of faculty of color, particularly Black scholars
Practice And Policy	Work alongside communities to develop campaigns that build trust and access to reliable and accurate information, education, and care	Support and develop on-the-ground work, mobile units, pop-up clinics, and transportation to services
		Engage and invest in community health workers
	Take collective action to redress inequities through structural change	Endorse equity-focused policy advocacy activities to address the synergistic conditions of racism and disease
		Advocate for voting rights protection, support policies that protect essential workers, expand insurance coverage, initiate family leave/care policies

Scholarship

Conceptual and theoretical opportunities

The language of syndemics aligns well with key social work concepts. Three outstanding areas of convergence and potential synergy are: intersectionality, vulnerability, and chronicity. Intersectionality gives insight into the origins of syndemics, how they are distributed, and experienced (Ferlatte et al., 2018). With increasing attention to intersectionality theory in social work (see, Azhar & Gunn, 2021; Matsuzaka et al., 2021), especially the experiences of multiply marginalized social groups and the ways that oppressive structures overlap and are interdependent, the syndemic concept offers a compatible frame for characterizing the complex, multi-level experiences and structures that intervene into equitable opportunities and outcomes. Similar to a core intersectionality argument, syndemic theory also resists an additive or otherwise arithmetic model of the relationships between disease and environment, stressing the synergistic interactions and complex, nuanced manifestations of those interactions.

Vulnerability is another key concept in social work scholarship that is also central to syndemic theory. Singer et al. (2017) described syndemic vulnerability as the "integration of epidemiological and experiential levels of analysis" of overlapping and clustering social and health conditions (p. 942). Syndemic vulnerability, therefore, requires an analysis of biopolitical vulnerability, the co-occurrence of biological vulnerability (e.g., increased allostatic load due to stresses of racism) and social/political vulnerability (e.g., SDOH; Ostrach & Singer, 2012). Of interest to social work scholars, and conceptually aligning with intersectionality, syndemic theory also asks for attention to structural vulnerability, patterns of suffering based on discrimination and exploitation, with particular emphasis on positionality (Quesada et al., 2011). Singer et al. (2017) also suggest the term *countersyndemics* to describe the protective benefit of certain individual traits, behaviors, or social conditions. This concept may serve to shed light on and give language to the self-protective, self-preserving nature of white supremacy at individual, social, and structural levels, particularly in the context of public health interventions addressing disproportionate disease burden, medical mistrust, and barriers to care.

Finally, at the conceptual level, syndemic theory may be useful to social work scholars focused on chronicity (Weaver & Mendenhall, 2014). The body of literature on chronicity is generally concerned with life course perspectives, the experiential level of analysis, meaning-making, and life continuity. The syndemic of racism and COVID-19, for example, may mark a critical juncture or turning point in the life course, particularly in terms of a shift in identity, future orientation, and the development of critical consciousness. These will be important topics for future social work scholarship.

Research opportunities

As discussed in length elsewhere (see for example, Hudson & Mehrotra, 2021), the syndemic of racism and COVID-19 presses on social work scholars to think and do differently in light of emerging social contexts and issues. The dynamic, multi-level, multi-system assumptions inherent to a syndemic frame require researchers to attend to, in particular: the integration of multiple levels of analysis, from experiential to epidemiological, and individual to ideological; the methodological requirements of these levels and types of analysis; and the overall impact they intend to have, and whether that impact is primarily academic or action-oriented.

Even within social work, research has often failed to reflect the lived experiences and meet the needs and expectations of minoritized racial/ethnic groups. To bridge this gap, there is also a need for greater community engagement in determining research and program priorities that are relevant to minoritized communities. Community-based participatory research (CBPR) is a useful approach for promoting health and reducing disparities, specifically by engaging communities affected by these negative outcomes in the research process. CBPR promotes ethical conduct of research by supporting collaborative partnerships between community-based, clinical/services settings and research institutions to improve health and well-being through action (Pinto et al., 2011). CBPR recognizes the strengths and experiences of *all* partners and attempts to shift the power of decision-making and knowledge production to community members who have been most affected by the issue at hand.

Additionally, to elevate voices of those who have been traditionally, systematically excluded, we need to build awareness of how racial, class, and gender positionalities (among others) shape assumptions and worldviews; challenge traditional power differentials in research relationships; and generate solutions that diverge from normative (often white, positivist) ways of conceptualizing, executing, and understanding research (Grégoire & Yee, 2007). Specifically, researchers should practice reflexivity and articulate the epistemological assumptions that underly chosen methodologies. There is also a need for openness in challenging traditional forms of positivist research paradigms in favor of those that engage in critical consciousness-raising (Grégoire & Yee, 2007). Such forms of scholarship are often more challenging to publish within social work journals, as the field tends to focus on empirical research on practice-oriented interventions. Moreover, editorial boards on which social work scholars serve ought to acknowledge that reviewers may critique and scrutinize work on certain topics, such as racism, more harshly than other submissions written by or presumed to be written by Black authors (Bell et al., 2021). Acknowledgment of this bias may serve to open up publication opportunities and expand critical scholarship in social work.

Applying a syndemic frame also informs the need for race-stratified data, and particularly the disaggregation of racial/ethnic data within Black, Indigenous, Latinx, and Asian communities, to reveal disparities addressed in clinical care and research. Social work researchers ought to analyze publicly available data on relationships between SDOH and epidemics, including COVID-19 disparities, which may be used to target specific communities at heightened risk for morbidity and mortality, focusing on population-level issues. At the state and local level, coordination may occur between public health departments, public housing authorities, welfare agencies and unemployment agencies to determine which census tracts, zip codes, or neighborhoods are at greatest risk for COVID-19 (Tipirneni, 2021) and develop specialized prevention strategies to target these communities while also considering community-identified priorities for program planning. Data must be presented with contextual factors that avoid placing the blame on minoritized groups for structural and social factors that are outside of their control, thereby avoiding biological explanatory models for health inequities (Nana-Sinkam et al., 2021).

Education opportunities

It is important for the profession to reflect on and work toward resolving its often-contradictory relationship to racial justice, beginning with revising the training and orientation within social work education. Social workers need to be informed of the racist legacies of the profession. For example, social work history often centers on charity organizations and the Settlement House movement, models which relied on systems of philanthropy and distancing between white, upper class "professionals" and the largely immigrant communities whom they served (Hall, 2021). Simultaneously, rich histories of mutual aid organizations and pioneers of color have been largely absent from social work history, further fortifying the mythology of social work as a predominantly white profession.

To rid itself of its explicit "Jane Addams canon," we must include the work of scholars who challenge this hegemonic discourse, such as Carlton-LaNey (e.g., 2001), Bent-Goodley (e.g., 2006), and Hounmenou (e.g., 2012). Educational practices within social work schools at the turn of the 19th century derived from the values, culture, and interests of affluent white women (Bowles et al., 2016; Carlton-LaNey, 1999) and arguably these traditions are still strongly upheld within social work education today. Ortega-Williams & McLane-Davidson, 2021, p. 566) describe this need for a revision of social work paradigms as "wringing out the whitewash," which metaphorically captures the labor of interrupting the Eurocentric epistemological hegemony underlying the social work canon. To address this biased orientation in social work education, we need curricular revisions and a resituating of social work pedagogy, including education about social workers of color, an understanding regarding the impacts of historical trauma, and engagement with critical, reconciliatory and Black liberation pedagogies (McCleary & Simard, 2021; Ortega-Williams & McLane-Davidson, 2021).

There are a number of strategies for social work education to move toward becoming antiracist, including infusing anti-oppressive content into course assignments, developing and integrating stand-alone courses on these topics, and thinking creatively about potential field placements. To support this work, best practices in antiracism pedagogy are urgently needed; in a recent systematic review of 150 studies in social work research addressing antiracism, published between 2008 and 2018, none of the reviewed studies focused on assessing antiracism as a learning outcome (Copeland & Ross, 2021). With little evidence to guide antiracist educational strategies, meaningful integration of anti-oppressive content will rely largely on the judgment and expertise of individual instructors and schools of social work. Further, the death of evidence and best educational practices makes it difficult to gauge how impactful these antiracist efforts may be in shaping the knowledge and skills of emerging social work professionals. Encouraging the development and evaluation of antiracist content in social work education should therefore be a priority for future social work research efforts.

For social work faculty, criteria for promotion and tenure, a system that fails to nurture and protect faculty of color, particularly Black faculty, have become more demanding while access to mentorship, protected time, and resources remain scarce (Bell et al., 2021). Faculty of color carry greater burdens in the forms of service on committees, mentorship of students of color, and participation in training and events centered on diversity, equity and inclusion (Azhar & DeLoach McCutcheon, 2021). To counteract these unbalanced trends, faculty of color need greater resources in the form of mentorship and support to facilitate their success within predominantly white institutions in academia. Together, these strategies would represent greater investment in the creation of a pipeline of future social workers and social work scholars of color.

Practice and policy opportunities

Syndemic theory requires social workers to consider what racial justice looks like across contexts and across practice and policy levels. Practice and policy implications include efforts to grow social work and health care workforce pipelines and various social policy opportunities to support advancement through a racial justice lens. For example, social work can revisit its community organizing roots by supporting, participating in and working alongside communities to develop campaigns that build trust and access to reliable and accurate health information, education, and care (Corbie-Smith, 2021). A syndemic frame requires attention to community-level initiatives that promote racial justice through a focus on "high touch" rather than "high tech" approaches to engagement: through on-the-ground work, mobile units, pop-up clinics, and transportation to services. Community Health Workers (CHWs) who are supervised by licensed social workers and registered nurses are known to fill such gaps in access through activities such as insurance enrollment, offering social support, engaging in health promotion initiatives, making referrals to social services, and assisting in accessing welfare services (e.g., Medicaid, SNAP, Section 8 housing vouchers; Rahman et al., 2021). CHWs are on the ground and have deep knowledge of the communities they serve as they tend to share a similar lived experience, culture, language, racial identity, neighborhood and socioeconomic needs as the clients they serve (Rahman et al., 2021). CHWs have the experiential knowledge to reckon an understanding of the relationships between lived experiences, health beliefs, and perceptions of truth among communities. Consequently, CHWs create relevance to scientific investigation and reporting, program development, and prevention programs that meet the unique needs of their community with varying levels of health literacy. As members of the communities they serve, CHWs are more easily able to develop rapport with community members and have the potential to reinstate trust in the medical community by offering critical public health and social service information. Health systems, local governments and state public health officials should continuously engage and invest in CHWs. Social work education should also prepare social workers to supervise such health care providers as another attempt to address the complex conditions of syndemics.

In addition to prioritizing community-driven initiatives and collaborations, the social work field must develop functional, anti-racist approaches that address the challenges within the profession itself. New evidence from a national survey of social workers conducted during the COVID-19 pandemic provides data to support a disturbing yet unsurprising trend of social workers of color experiencing elevated levels of physical and mental health concerns, as well as concerns related to PPE access, relative to their white counterparts (Ross et al., 2022). Applying a syndemic frame to eliminate these problematic population-level trends within the profession itself requires both profession-specific efforts that prioritize the mental, physical, and financial health and well-being of social workers of color in combination with larger-scale policy changes that redress longstanding SDOH inequities that pervade systems and institutions across the country (Ross et al., 2022).

Social and health policy implications

As Godlee (2020) summarized, syndemic theory attends to "large-scale, political-economic forces, which play out over generations, result[ing] in deep-seated social, economic, and power inequities" that "shape the distribution of risks and resources for health" (p. 2). Corresponding interactions between disease and the SDOH need to be addressed through structural policy and institutional interventions at the local, national, and global levels (World Health Organization [WHO], 2019). A syndemic frame can help guide equity-focused policy advocacy activities to address the synergistic conditions of racism and disease across the SDOH and PEH (Shim & Starks, 2021). For example, voting rights protection policies, policies that protect essential workers, expanding insurance coverage and income and family support policies all take a syndemic approach, by accounting for the intersecting and synergistic conditions of social, economic, and environmental injustice that communities of color face. While vaccination in the U.S. continues to ensue, the syndemic of racism and COVID-19 is far from over. It is likely that communities of color will continue to experience pronounced, disproportionate effects of COVID-19 and its sequelae for decades to come, perhaps longer if structural inequities fueled by systemic racism remain unacknowledged and unaddressed. While a syndemic frame provides the social work profession with an opportunity to "see the problem," more clearly (Crenshaw, 2016), our professional standards and the urgency of these syndemic conditions demand that we take collective action to redress inequities through large scale structural change. Social workers must advocate for and implement practice and policy solutions that account for social and economic contexts – including the SDOH and PEH and their intersectional impacts – if we are to truly advance health equity.

Disclosure statement

No potential conflict of interest was reported by the author(s).

ORCID

Kimberly D. Hudson (iD) http://orcid.org/0000-0001-9231-8312
Sameena Azhar (iD) http://orcid.org/0000-0002-2249-8976
Rahbel Rahman (iD) http://orcid.org/0000-0003-1065-7084
Abigail M. Ross (iD) http://orcid.org/0000-0003-3706-4166

References

Abrams, E. M., & Szefler, S. J. (2020). COVID-19 and the impact of social determinants of health. *The Lancet Respiratory Medicine*, 8(7), 659–661. https://doi.org/10.1016/S2213-2600(20)30234-4

American Public Health Association. (2021). *Racism is a public health crisis*. Retrieved April 7, 2021, from https://www. apha.org/topics-and-issues/health-equity/racism-and-health/racism-declarations

Anyane Yeboa, A., Sato, T., & Sakuraba, A. (2020). Racial disparities in COVID 19 deaths reveal harsh truths about structural inequality in America. *Journal of Internal Medicine*, 288(4), 479–480. https://doi.org/10.1111/joim.13117

Azhar, S., & DeLoach McCutcheon, K. (2021, September). How racism against BIPOC women faculty operates in social work academia. *Advances in Social Work*, 21(2/3), 396–420. https://doi.org/10.18060/24118

Azhar, S., & Gunn, A. (2021, August). Navigating intersectional stigma: Strategies for coping among cisgender women of color in two qualitative studies. *Qualitative Health Research*, 31(12), 2194–2210. https://doi.org/10.1177/10497323211025249

Bailey, Z. D., & Moon, J. R. (2020). Racism and the political economy of COVID-19: Will we continue to resurrect the past? *Journal of Health Politics, Policy and Law*, 45(6), 937–950. https://doi.org/10.1215/03616878-8641481

Bajaj, S. S., & Stanford, F. C. (2021). Beyond Tuskegee—Vaccine distrust and everyday racism. *New England Journal of Medicine*, 384(5), e12. https://doi.org/10.1056/NEJMpv2035827

Bedford, H., Attwell, K., Danchin, M., Marshall, H., Corben, P., & Leask, J. (2018). Vaccine hesitancy, refusal and access barriers: The need for clarity in terminology. *Vaccine*, 36(44), 6556–6558. https://doi.org/10.1016/j.vaccine.2017.08.004

Bell, M. P., Berry, D., Leopold, J., & Nkomo, S. (2021). Making Black lives matter in academia: A Black feminist call for collective action against anti-blackness in the academy. *Gender, Work, and Organization*, 28(S1), 39–57. https://doi.org/10.1111/gwao.12555

Bell, C. C., & Dunbar, E. (2012). Racism and pathological bias as a co-occurring problem in diagnosis and assessment. In T. A. Widiger (Ed.), *The Oxford Handbook of Personality Disorders* (pp. 694–709). Oxford University Press. https://doi.org/10.1093/oxfordhb/9780199735013.013.0032

Bent-Goodley, T. B. (2006). Oral histories of contemporary African American social work pioneers. *Journal of Teaching in Social Work*, 26(1–2), 181–199. https://doi.org/10.1300/J067v26n01_11

Bogart, L. M., Ojikutu, B. O., Tyagi, K., Klein, D. J., Mutchler, M. G., Dong, L., & Kellman, S. (2021). COVID-19 related medical mistrust, health impacts, and potential vaccine hesitancy among Black Americans living with HIV. *Journal of Acquired Immune Deficiency Syndromes (1999)*, 86(2), 200. https://doi.org/10.1097/QAI.0000000000002570

Bowles, D. D., Hopps, J. G., & Clayton, O. (2016). The impact and influence of HBCUs on the social work profession. *Journal of Social Work Education*, 52(1), 118–132. https://doi.org/10.1080/10437797.2016.1112650

Brodie, N., Perdomo, J. E., & Silberholz, E. A. (2021). The dual pandemics of COVID-19 and racism: Impact on early childhood development and implications for physicians. *Current Opinion in Pediatrics*, 33(1), 159–169. https://doi.org/10.1097/MOP.0000000000000985

Buchanan, L., Bui, Q., & Patel, J. (2020). *Black Lives Matter may be the largest movement in the U.S. history*. New York Times. Retrieved May 25, 2021, from https://www.nytimes.com/interactive/2020/07/03/us/george-floyd-protests-crowd-size.html

Bunch, L. (2021, June). A tale of two crises: Addressing Covid-19 vaccine hesitancy as promoting racial justice. *HEC Forum*, 33(1–2), 143–154. https://doi.org/10.1007/s10730-021-09440-0

Carlton-LaNey, I. (1999). African American social work pioneers' response to need. *Social Work*, 44(4), 311–321. https://doi.org/10.1093/sw/44.4.311

Carlton-LaNey, I. (Ed.). (2001). *African American leadership: An empowerment tradition in social welfare history*. NASW Press.

Centers for Disease Control and Prevention. (2021a). *COVID data tracker*. https://covid.cdc.gov/covid-data-tracker/#datatracker-home

Centers for Disease Control and Prevention. (2021b). *Estimated disease burden of COVID-19*. https://www.cdc.gov/coronavirus/2019-ncov/cases-updates/burden.html

Centers for Disease Control and Prevention. (2021c). *CDC COVID-19 vaccination program provider requirements and support*. https://www.cdc.gov/vaccines/covid-19/vaccination-provider-support.html

Copeland, P., & Ross, A. (2021). Assessing antiracism as a learning outcome in social work education: A systematic review. *Advances in Social Work*, 21(2/3), 766–778. https://doi.org/10.18060/24139

Corbie-Smith, G. (2021, March). Vaccine hesitancy is a scapegoat for structural racism. In *JAMA Health Forum* (Vol. 2, No. (3), pp. e210434–e210434). American Medical Association.

Crenshaw, K. (2016, October). *The urgency of intersectionality* [Video]. https://www.ted.com/talks/kimberle_crenshaw_the_urgency_of_intersectionality

DeBruin, D., Liaschenko, J., & Marshall, M. F. (2012). Social justice in pandemic preparedness. *American Journal of Public Health*, 102(4), 586–591. https://doi.org/10.2105/AJPH.2011.300483

Devakumar, D., Selvarajah, S., Shannon, G., Muraya, K., Lasoye, S., Corona, S., Paradies, Y., Abubakar, I., & Achiume, E. T. (2020). Racism, the public health crisis we can no longer ignore. *The Lancet*, 395(10242), e112–e113. https://doi.org/10.1016/S0140-6736(20)31371-4

Dobbins, J. E., & Skillings, J. H. (2000). Racism as a clinical syndrome. *American Journal of Orthopsychiatry*, 70(1), 14–27. https://doi.org/10.1037/h0087702

Ferlatte, O., Salway, T., Trussler, T., Oliffe, J. L., & Gilbert, M. (2018). Combining intersectionality and syndemic theory to advance understandings of health inequities among Canadian gay, bisexual and other men who have sex with men. *Critical Public Health*, *28*(5), 509–521. https://doi.org/10.1080/09581596.2017.1380298

Godlee, F. (2020). Racism: The other pandemic. *BMJ*, 369. https://doi.org/10.1136/bmj.m2303

Gravlee, C. C. (2020). Systemic racism, chronic health inequities, and COVID -19: A syndemic in the making? *American Journal of Human Biology*, *32*(5), 1–8. https://doi.org/10.1002/ajhb.23482

Grégoire, H., & Yee, J. Y. (2007). Ethics in community-university partnerships involving racial minorities: An anti-racism standpoint in community-based participatory research. In *Partnership perspectives IV* (Vol. I, pp. 70–79). Community-Campus Partnerships for Health.

Hall, R. E. (2021). Social work's feminist Façade: Descriptive manifestations of white supremacy. *The British Journal of Social Work*, *52*(2), 1055–1069. https://doi.org/10.1093/bjsw/bcab093

Harvey, M. (2021). The political economy of health: Revisiting its Marxian origins to address 21st-century health inequalities. *American Journal of Public Health*, *111*(2), 293–300.

Horton, R. (2020). Offline: COVID-19 is not a pandemic. *The Lancet*, *396*(10255), 874. https://doi.org/10.1016/S0140-6736(20)32000-6

Hounmenou, C. (2012). Black settlement houses and oppositional consciousness. *Journal of Black Studies*, *43*(6), 646–666. https://doi.org/10.1177/0021934712441203

Hudson, K. D., & Mehrotra, G. R. (2021). Pandemic and protest in 2020: Questions and considerations for social work research. *Qualitative Social Work*, *20*(1–2), 264–270. https://doi.org/10.1177/1473325020973315

Johnson-Agbakwu, C. E., Ali, N. S., Oxford, C. M., Wingo, S., Manin, E., & Coonrod, D. V. (2022). Racism, COVID-19, and health inequity in the USA: A call to action. *Journal of Racial and Ethnic Health Disparities*, *9*, 52–58. https://doi.org/10.1007/s40615-020-00928-y

Kahn, J. (2017). Pills for prejudice: Implicit bias and technical fix for racism. *American Journal of Law & Medicine*, *43*(2–3), 263–278. https://doi.org/10.1177/0098858817723664

Kaiser Family Foundation. (2021, April 28). *Latest data on COVID-19 vaccinations race/ethnicity*. https://www.kff.org/coronavirus-covid-19/issue-brief/latest-data-on-covid-19-vaccinations-race-ethnicity/

Khosla, N. N., Perry, S. P., Moss-Racusin, C. A., Burke, S. E., & Dovidio, J. F. (2018). A comparison of clinicians' racial biases in the United States and France. *Social Science & Medicine*, *206*, 31–37. https://doi.org/10.1016/j.socscimed.2018.03.044

Krieger, N. (2020). ENOUGH: COVID-19, structural racism, police brutality, plutocracy, climate change—and time for health justice, democratic governance, and an equitable, sustainable future. *American Journal of Public Health*, *110*(11), 1620–1623. https://doi.org/10.2105/AJPH.2020.305886

Krishnan, L., Ogunwole, S. M., & Cooper, L. A. (2020). Historical insights on coronavirus disease 2019 (COVID-19), the 1918 influenza pandemic, and racial disparities: Illuminating a path forward. *Annals of Internal Medicine*, *173*(6), 474–481. https://doi.org/10.7326/M20-2223

Laurencin, C. T., & McClinton, A. (2020). The COVID-19 pandemic: A call to action to identify and address racial and ethnic disparities. *Journal of Racial and Ethnic Health Disparities*, *7*(3), 398–402. https://doi.org/10.1007/s40615-020-00756-0

Laurencin, C. T., & Walker, J. M. (2020). A pandemic on a pandemic: Racism and COVID-19 in Blacks. *Cell Systems*, *11*(1), 9–10. https://doi.org/10.1016/j.cels.2020.07.002

Leitch, S., Corbin, J. H., Boston-Fisher, N., Ayele, C., Delobelle, P., Gwanzura Ottemöller, F., . . . & Wicker, J. (2020). Black lives matter in health promotion: Moving from unspoken to outspoken. *Health Promotion International*, *36*(4), 1160–1169. https://doi.org/10.1093/heapro/daaa121

Marmot, M., & Wilkinson, R. (Eds.). (2005). *Social determinants of health*. Oxford University Press.

Martínez, M. E., Nodora, J. N., & Carvajal Carmona, L. G. (2021). The dual pandemic of COVID 19 and systemic inequities in US Latino communities. *Cancer*, *127*(10), 1548–1550. https://doi.org/10.1002/cncr.33401

Matsuzaka, S., Hudson, K. D., & Ross, A. M. (2021). Operationalizing intersectionality in social work research: Approaches and limitations. *Social Work Research*, *45*(3), 155–168. https://doi.org/10.1093/swr/svab010

McCleary, J., & Simard, E. (2021). Honoring our ancestors: Using reconciliatory pedagogy to dismantle white supremacy. *Advances in Social Work*, *21*(2/3), 259–273. https://doi.org/10.18060/24146

McCoy, H. (2020). Black lives matter, and yes, you are racist: The parallelism of the twentieth and twenty-first centuries. *Child and Adolescent Social Work Journal*, *37*(5), 463–475. https://doi.org/10.1007/s10560-020-00690-4

Mendenhall, E. (2020). The COVID-19 syndemic is not global: Context matters. *The Lancet*, *396*(10264), 1731. https://doi.org/10.1016/S0140-6736(20)32218-2

Nana-Sinkam, P., Kraschnewski, J., Sacco, R., Chavez, J., Fouad, M., Gal, T., & Behar-Zusman, V. (2021). Health disparities and equity in the era of COVID-19. *Journal of Clinical and Translational Science*, *5*(e99), 1–8. https://doi.org/10.1017/cts.2021.23

Office of Disease Prevention and Health Promotion. (2018). *Maternal, infant, and child health*. Healthy People 2020. U.S. Department of Health and Human Services. https://www.healthypeople.gov/2020/topics-objectives/topic/maternal-infant-and-child-health/objectives

Økland, H., & Mamelund, S.-E. (2019). Race and 1918 influenza pandemic in the United States: A review of the literature. *International Journal of Environmental Research and Public Health, 16*(14), 2487. https://doi.org/10.3390/ijerph16142487

Ortega-Williams, A., & McLane Davison, D. (2021). Wringing Out the "Whitewash": Confronting the Hegemonic Epistemologies of Social Work Canons (Disrupting the Reproduction of White Normative). *Advances in Social Work, 21*(2/3), 566–587. https://doi.org/10.18060/24475

Ostrach, B., & Singer, M. (2012). At special risk: Biopolitical vulnerability and HIV/STI syndemics among women. *Health Sociology Review, 21*(3), 258–271. https://doi.org/10.5172/hesr.2012.21.3.258

Pinto, R. M., Spector, A. Y., & Valera, P. A. (2011). Exploring group dynamics for integrating scientific and experiential knowledge in community advisory boards for HIV research. *AIDS Care, 23*(8), 1006–1013. https://doi.org/10.1080/09540121.2010.542126

Poussaint, A. (1999, August 26). They hate. They kill. Are they insane? [Op-Ed piece]. *New York Times.*

Quesada, J., Hart, L. K., & Bourgois, P. (2011). Structural vulnerability and health: Latino migrant laborers in the United States. *Medical Anthropology, 30*(4), 339–362. https://doi.org/10.1080/01459740.2011.576725

Rahman, R., Ross, A., & Pinto, R. (2021). The critical importance of community health workers as first responders to COVID-19 in USA. *Health Promotion International, 36*(1), 1–10. https://doi.org/10.1093/heapro/daaa069

Ross, A. M., Cederbaum, J. A., de Saxe Zerden, L., Zelnick, J. R., Ruth, B. J., & Guan, T. (2022). Bearing a disproportionate burden: Racial/ethnic disparities in experiences of US-based social workers during the COVID-19 pandemic. *Social Work, 67*(1), 28–40. https://doi.org/10.1093/sw/swab050

Shim, R. S., & Starks, S. M. (2021). COVID-19, structural racism, and mental health inequities: Policy implications for an emerging syndemic. *Psychiatric Services, 72*(10), 1193–1198. https://doi.org/10.1176/appi.ps.202000725

Singer, M. (2009). *Introducing Syndemics: A critical systems approach to public and community health.* Jossey-Bass.

Singer, M., Bulled, N., Ostrach, B., & Mendenhall, E. (2017). Syndemics and the biosocial conception of health. *The Lancet, 389*(10072), 941–950. https://doi.org/10.1016/S0140-6736(17)30003-X

Skillings, J. H., & Dobbins, J. E. (1991). Racism as a disease: Etiology and treatment implications. *Journal of Counseling & Development, 70*(1), 206–212. https://doi.org/10.1002/j.1556-6676.1991.tb01585.x

Tan, S. B., DeSouza, P., & Raifman, M. (2021). Structural racism and COVID-19 in the USA: A county-level empirical analysis. *Journal of Racial and Ethnic Health Disparities, 8*(1), 1–11. https://doi.org/10.1007/s40615-020-00905-5

Thakur, N., Lovinsky-Desir, S., Bime, C., Wisnivesky, J. P., & Celedón, J. C. (2020). The structural and social determinants of the racial/ethnic disparities in the US COVID-19 pandemic. What's our role? *American Journal of Respiratory and Critical Care Medicine, 202*(7), 943–949. https://doi.org/10.1164/rccm.202005-1523PP

Thomas, J. M. (2014). Medicalizing racism. *Contexts, 13*(4), 24–29. https://doi.org/10.1177/1536504214558213

Thomas, J. M., & Byrd, W. C. (2016). The "sick" racist. *Du Bois Review, 13*(1), 181. https://doi.org/10.1017/S1742058X16000023

Tipirneni, R. (2021). A data-informed approach to targeting social determinants of health as the root causes of COVID-19 disparities. *American Journal of Public Health, 111*(4), 620–622. https://doi.org/10.2105/AJPH.2020.306085

Weaver, L. J., & Mendenhall, E. (2014). Applying syndemics and chronicity: Interpretations from studies of poverty, depression, and diabetes. *Medical Anthropology, 33*(2), 92–108. https://doi.org/10.1080/01459740.2013.808637

Wellman, D. (2000). From evil to illness: Medicalizing racism. *American Journal of Orthopsychiatry, 70*(1), 28–32. https://doi.org/10.1037/h0087669

World Bank. (2018). *Current health expenditure per capita.* https://data.worldbank.org/indicator/SH.XPD.CHEX.PC.CD

World Health Organization. (2019). *Social determinants of health.* https://www.who.int/social_determinants/en/

Visualizing structural competency: moving beyond cultural competence/ humility toward eliminating racism

Eric Kyere, Stephanie Boddie and Jessica, Euna Lee

ABSTRACT

In this article, the authors argue that in the United States, structural racism set the stage that increased persons of color's vulnerabilities and risks to COVID-19 compared to Whites, while simultaneously killing Blacks through racialized policing. They draw on structural violence as a theoretical framework to ground their argument and add to the discussion on the need for social work to explicitly build structural competency to effectively respond to structural racism. Most importantly, the authors contend that, structural racism entails a network of interdependent institutions and organizations that interact with individuals in a complex way to affect health and well-being. Therefore, eliminating racism needs to move beyond a single institution and organization to interdependent relationships among institutions and the mechanized paths through which their effects are translated at the community and individual levels. In this regard, instead of simplifying the complexities surrounding structural racism, we should embrace them and build knowledge system and tools that are complexity sensitive toward eliminating racism. The authors extend the emerging discussion on a renewed focus for structural competency in social work education and respond to the Grand Challenge to Eliminate Racism by presenting a "structuragram" as a heuristic to assess, analyze, and intervene at the structural level factors that influence the individual and community's realities. We conclude with a case example and recommendations for structural competency-based practice.

Introduction

In the context of persistent racial inequities in health and well-being that disproportionately disadvantage certain groups and communities, this article contributes to and extends the emerging search for structural competency in social work education, practice, and research. Social work, which originally started with a focus on the person-in-environment (PIE) framework to understand and intervene in social problems, can be a key player in shaping how structural forces contribute to the causes, intervention, and prevention of social conditions that affect health and well-being. The newest Social Work Grand Challenge recognizes the complexity of racial inequality in the US and the need for change at the individual, organizational, community, and societal levels (Teasley et al., 2021). More recently, some social workers and other health and helping professionals have called for more attention to structural forces (e.g., social policies and structural racism) to address inequities that shape health and well-being and structural competency in training, research, and practice (Gee & Hicken, 2021; Neff et al., 2020; Sacks & Jacobs, 2019; Wills, 2021). Eliminating racism must move beyond a single institution to interdependent relationships among institutions and the pathways through which their effects are translated at the community and individual levels (Gee & Hicken, 2021).

In this article, we contribute to the discussion on structural competency – the ability to understand interdependent relationships among social institutions (e.g., health care, law enforcement, education, housing, media, tax system, the market) – and the mechanisms (e.g., policies, processes, and practices) by which networks of institutions operate to translate effects that surround individual-level conditions (Booth, 2020; J. M. Metzl & Hansen, 2021). We describe how structural competency moves beyond cultural competency and cultural humility to understand the social, political, and economic environments over time that contribute to disparate outcomes for people of color. In doing this, we draw on structural violence, as a theoretical framework, to highlight the connection between structural racism and COVID-19 in the US and strengthen our call for structural competency in social work. We contend that structural competency will enhance understanding of structural racism as a form of structural violence that sets the stage for the disproportionate impact of COVID-19 on people of color in the US, especially Black people, alongside its longstanding legitimization of police brutality and unjust killings. We introduce a conceptual model – *structuragram* – as a heuristic to build structural-based competency for social work assessment, analysis, and intervention. We conclude with a case example and recommendations for structural competency-based practice.

From cultural competence to structural competence

Although not new to social work, an explicit focus on structural competency is an emerging paradigm in healthcare education that attends to the network of political, social, economic, environmental, and other structural forces and their effects on individual and group behaviors relative to health and well-being (Downey et al., 2019; Jacobs & Mark, 2019; J. M. Metzl & Hansen, 2021). Over the past several decades, social work has stressed the need for cultural competency – striving for mastery – to effectively serve the diverse client system (Kohli et al., 2010). Additionally, cultural humility, an interpersonal and client-centered stance that requires ongoing learning, has also become prominent in social work education (Fisher, 2021; Fisher-Borne et al., 2015). While these concepts have motivated the social work profession to develop strategies to meaningfully engage with diverse populations, they are limited in establishing how the cultural expression of identity is constrained by structural forces that control and regulate material and socially enhancing resources affecting health and well-being (Chambers & Ratliff, 2019; Jacobs & Mark, 2019). Consequently, in the US, individuals, groups, and communities (e.g., racial, and ethnic minorities, immigrants, women, and sexual minorities) whose identity and cultural means of expression diverge from those of the socially dominant groups (e.g., white, male, heterosexual, citizens) have restricted access to opportunities to enhance their health and well-being.

Structural competency moves beyond cultural competence and cultural humility and can more effectively address the social determinants of health (SDoH). SDoH are the conditions in which people are born, grow, live, play, work, worship, age, and the broader systems that account for most health outcomes (World Health Organization (WHO), 2021). The causal pathways of SDoH are complex and consist of upstream, midstream, and downstream factors, each of which can be evaluated to contribute to our understanding of disparate health outcomes (World Health Organization (WHO), 2021). Upstream factors are the root causes and conditions contributing to health disparities, such as social policies and institutions (e.g., war on drugs policies, Health and Human Services; Robichaux & Sauerland, 2021). Midstream factors comprise physical and social environments, including the resources that characterize them (e.g., healthcare services, schools, housing, food access, water quality within neighborhoods, medical mistrust/provider bias). These factors are critical to determining how upstream factors translate at the interpersonal level and the recursive relationships between upstream and downstream determinants (Booth, 2020; Williams et al., 2019b). Downstream factors are the most observable as they manifest at the individual level (e.g., health behaviors, diseases, injuries, deaths, interventions; Braveman & Gottlieb, 2014; Cassels & van den Abbeele, 2021). More efforts to address upstream factors are needed because they are fundamental to identifying the fundamental causes of inequities (Braveman et al., 2011).

Structural competency from an upstream perspective includes awareness of structural conditions and entails structural responsiveness – that is, how structural factors are historically grounded and produce contemporary effects as well as the pathways by which such effects are translated from structural roots to individual experiences of health and well-being. Many existing tools, models, and interventions address downstream factors and even some midstream factors, but what is missing are approaches to comprehensive assessment that address which upstream factors at the structural level facilitate health and well-being via midstream factors. J. M. Metzl and Hansen (2021) adapted a poststructuralist conceptualization of "structure," one which includes invisible and visible elements. Tangible structures consist of institutions, the built environment, public infrastructure, and communication systems, among other arrangements. Intangible structures may involve language, discourse, bureaucratic frameworks, ideologies, and stigma. Booth (2020) expanded on J. M. Metzl and Hansen's (2021) conceptualization of structural competency by integrating Giddens' theory of the bi-directional relationship between structures and individuals and highlighting the concept of agency and opportunities for individuals to change social structures. Booth (2020), like J. M. Metzl and Hansen (2021), also acknowledged criticisms of structuration theory as being ahistorical and a priori. In our model, we attend to the historical roots of structural forces that anchor the unjust contexts surrounding inequities in health and well-being.

Structural violence

Structural competency highlights the concepts of structural violence and the denaturalization of social hierarchies inherent in structural racism (Neff et al., 2020). In this section, we draw on structural violence as a theoretical framework to argue that structural racism is a form of violence that can help elucidate the disparate impact of COVID-19 on marginalized groups and the racialized killings of persons of color, especially Black people. According to Johan Galtung (1969), structural violence explains the gap between what could be, like higher life expectancy, and what actually exists, like premature death: "It is that which increases the distance between the potential and the actual, and that which impedes the decrease of this distance" to create the absence of peace physically, psychologically, socially, and spiritually (Galtung, 1969, p. 168). Structural violence is a historically rooted and resource-driven process and structure that restricts agency and generates suffering for certain groups and individuals through the unequal distribution of power and resources (Farmer, 1996; Galtung, 1969). This violence is different from the destruction of life and well-being due to disasters or events beyond human control. Structural violence is evident when a social system restricts the availability, accessibility, and use of resources (e.g., health care, education, housing, employment, food systems, and the credit market) needed to meet or even exceed basic needs, thereby potentially maximizing well-being (R. B. Reich, 2020b; Galtung, 1969). According to Laurie and Shaw (2018), structural violence results in a truncated life, which is:

> life that is robbed of its potential: life that is forcibly "humiliated, ashamed, anxious, harassed, stigmatized, and depressed" … In all these cases, violent conditions that warp and destroy self-realization and actualization. They preserve the gulf between "what is, and what could have been", conspiring with the conditioning that keeps people in their place: mentally, physically, economically, and culturally. (Laurie & Shaw, 2018, p. 12)

Ultimately, structural violence demonstrates how unjust social structures lead to both bodily and emotional suffering, including premature death, over-representation in stigmatizing institutions, or states of inequality and misery, among marginalized individuals and communities (Farmer, 1996; Neff et al., 2020).

Like SDoH, structural violence highlights avoidable conditions that culminate in disparate disadvantages for marginalized groups. Although the impact of structural violence is felt at the micro-levels of interpersonal interactions, it is anchored in upstream-level forces or in the political realm (R. B. Reich, 2020b; Laurie & Shaw, 2018). The result is a deficit-based context and constrained access to resources needed to meet basic needs for survival, well-being, identity, and freedom (Galtung, 1969). Structural violence primarily operates through exclusion, division, exploitation, oppression, and humiliation, perpetuating conditions in which those who are socially powerful have their needs met more often and their interests more fully satisfied from the system of capitalism compared to those who are marginalized and consequently disadvantaged (Galtung, 1969; Laurie & Shaw, 2018; R. B. Reich, 2020b; Farmer, 1996). While not deadly, structural violence "works on the soul" and endures through "lies, brainwashing, indoctrination of various kinds, etc. that serve to decrease mental potentialities" (Galtung, 1969, p. 170). However, Farmer (1996) suggested four ways through which structural violence may generate direct violence:

1. Structurally subordinated groups are likely to use direct forms of violence, such as rioting, to rise up against existing structures.
2. Those maintaining or benefiting from existing structures are more likely to use violence through police or military enforcement to keep the peace and preserve the status quo.
3. In communities lacking resources due to structural practices, competition for limited resources can also lead to violence between groups.
4. Some subordinated groups are vulnerable to scapegoating or inter-group conflict.

Those living under the thumb of oppressive structures may also become complicit in perpetuating the system. For most people, structural violence is difficult to recognize. It appears as a normal, inevitable, and accepted part of everyday life or is dismissed as simply the way the world operates (Galtung, 1969). Structural violence is also often downplayed or overlooked until it escalates to more explicit forms of violence (e.g., murder, domestic violence, robbery, bullying, emotional manipulation, and even apathy; Finlev, 2012; Galtung, 1969). COVID-19 has more clearly exposed some of the structural patterns that produce inequalities in wealth, power, and access to resources and services that in turn cultivate the conditions under which differential outcomes in health and well-being, like premature death, occur (Singer & Rylko-Bauer, 2020; Williams et al., 2019a).

The violence of structural racism: COVID-19 and police brutality

The global COVID-19 pandemic has unevenly impacted the global community and continues to do so in many ways. Persons of color (e.g., Blacks, Native Americans, and Hispanics) and other marginalized groups and communities (individuals in prisons, those in-home care, those experiencing homelessness) have been severely impacted, economically, and psychologically, by the imposition of pandemic mitigation measures (Singer & Rylko-Bauer, 2020). These groups and communities have also experienced higher COVID-19 morbidity and mortality rates (Czeisler et al., 2020; R. Reich, 2020a; Van Dorn et al., 2020). At the same time, racialized forms of terrorism, including police brutality and killings of Black people and other persons of color, have continued unabated (Bavel et al., 2020; Blankenship & Reeves, 2020; Cohen, 2020). A review of research (Alexander, 2020; Bailey et al., 2021, 2017; Jacoby et al., 2018; Williams et al., 2019a) linking racial disparities across systems (e.g., health care, education, housing, employment, criminal justice) suggests that the factors responsible for increased COVID-19 morbidity and mortality rates among people of color as well as the factors driving police brutality are both rooted in structural racism and are interconnected (Cassels & van den Abbeele, 2021; Gee & Hicken, 2021; Laster Pirtle, 2020).

We define structural racism as ideas and methods of domination and subordination that entail interconnected institutions (e.g., health care, education, housing, employment, law enforcement) and culturally reinforcing beliefs and practices with historical origins and designs that confer power and resources on some individuals or groups while stripping power and resources from others (Bailey et al., 2017; Virdee, 2019; Williams et al., 2019a). With particular attention to the inequities permeating the social and economic systems that shape our daily lives, we consider structural racism as structural violence. Ultimately, structural racism functions to produce and reproduce a social reality in which race has tremendous power to transform the ontological state of being, situating racial minorities and those experiencing poverty in socially murderous conditions that truncate potential and often result in premature death (Jacoby et al., 2018; Medvedyuk et al., 2021; Rosa & Díaz, 2019). As a result, persons of color, such as Black people, tend to be more susceptible to earlier onset of illness, more aggressive progression of illness, and poorer survival outcomes (Williams et al., 2019a). One mechanism that can help clarify how structural racism persists and affects the health and well-being of people of color is residential segregation (Jacoby et al., 2018; Williams et al., 2019a). Residential segregation was designed to ensure the separation of White communities from Black communities. Primarily driven by government policies, subsidies, and tax codes dating back to the 1800s as well as government support for private policies, such as restrictive covenants, redlining, discriminatory zoning, and mortgage discrimination, residential discrimination has been legitimized (Jacoby et al., 2018; Rothstein, 2017; Williams et al., 2019a). Residential segregation limits access by Black people and other persons of color to high-quality schools, optimal healthcare services, various employment opportunities, and better-quality food and places them in closer proximity to impoverished and crime-ridden neighborhoods as well as sites contaminated by toxins and other hazardous materials (Jacoby et al., 2018; Wilkerson, 2019b; Williams et al., 2019a). In such socially isolated communities, people of color are highly vulnerable to racialized policing and criminal justice institutions, resulting in hyper-incarceration rates, rampant police brutality, and homicides with consequent traumatic impacts across the life span (Alexander, 2020; Calathes, 2017; Smith Lee & Robinson, 2020).

By December 11, 2020, there had been around 15,474,800 confirmed cases of COVID-19 in the US, with 291,522 deaths (Richardson et al., 2020). When data on morbidity and mortality rates as well as on the impact of pandemic mitigation efforts first emerged, striking disparities became apparent. Persons of color, such as Blacks, Native Americans, Latinos, and Asians, immigrants and other individuals experiencing poverty and homelessness, essential workers, the aged, and incarcerated persons, were disproportionately impacted (Czeisler et al., 2020; Garg et al., 2020; R. Reich, 2020a; Van Dorn et al., 2020). The Centers for Disease Control and Prevention (CDC) reported that Blacks/African Americans were being infected with COVID-19 at higher rates than Whites and Hispanics in 14 states. For example, data from New York indicated that per 100,000 COVID-19-related deaths, 92.3 were Black/African Americans and 74.3 were Hispanic/Latino individuals compared to 45.2 for Whites and 34.5 for Asians (Centers for Disease Control and Prevention, 2020). Relatedly, researchers analyzed data from the *Epic* health record system, which included 7 million Black patients and 34.1 million White patients, collected by July 2020. They found that the hospitalization and death rates per 10,000 individuals were 24.6 and 5.6 for Blacks compared to 7.4 and 2.3 for Whites, respectively (Lopez et al., 2021). Similarly, Richardson et al. (2020) analyzed comparative data from South Korea and Louisiana with similar characteristics relative to population density and the GINI coefficient (the measure of income distribution across the population) to estimate whether living in high-density areas alone could help explain vulnerability to COVID-19 infection and its health outcomes. Their findings suggested that structural factors like overcrowded living conditions, increased exposure to frontline work, and limited access to preventive measures, rather than population density, explained the difference in health outcomes between respondents with similar income. They concluded that:

> [I]n the United States where the problem of the 21st century is still the problem of the color line, 400 years of structural racism, violently seized privilege, and continuous trauma from racial terror and dehumanization are clearly manifested in the disproportionate incidence and mortality rates of COVID-19 among Black Americans. (p. 3)

Historically, structural racism has functioned to isolate persons of color, especially Black people, into multigenerational households, crowded conditions, and low-wage jobs, such as nursing aids, transit workers, and grocery store clerks, that are impossible to perform remotely (Laster Pirtle, 2020; Lopez et al., 2021; R. Reich, 2020a). A review of historical evidence linking racism to health inequities (Kyere, 2022) suggested that structural racism through these mechanisms (e.g., crowded living conditions) directly increased the risk of COVID-19 infection in individuals and communities and indirectly through chronic conditions, such as diabetes, hypertension, obesity, and cardiovascular diseases (Jean-Baptiste & Green, 2020; Williams et al., 2019a).

Overall, a robust body of scholarship suggests that understanding racism's historical roots, structural design, and contemporary manifestation is important for comprehending the dynamics of violent conditions that enhance the risk of mortality among individuals, families, and communities (Bailey & Moon, 2020; Jacoby et al., 2018; O'Brien et al., 2020; Richardson et al., 2020; Williams et al., 2019b) in health crises such as the COVID-19 pandemic. For example, Deitz and Meehan (2019) examined household plumbing to identify which individuals are more likely to live in water-deprived communities. They found that living in a Black, Hispanic, or American Indian or Alaskan Native household increases the likelihood of experiencing water insecurity. Deitz and Meehan's research linked household water insecurity not to technical problems related to supply and engineering (downstream SDoH) but instead to racially designed institutional practices, such as residential segregation and disinvestment in communities of color. Living in areas without access to a safe and reliable water supply could undermine the ability of individuals and communities to adhere to COVID-19 mitigation measures, such as regular hand washing and household cleaning (Bailey & Moon, 2020), in turn increasing the risk of infection and, potentially, death. Dr. Abdul El-Sayed, former director of the city of Detroit Health Department, powerfully conveyed how the violence of structural racism made it possible for COVID-19 to spread rapidly among Black residents of Detroit. In an interview on Democracy Now, El-Sayed (2020) stated:

> When you look at communities that are suffering the most, they're communities on which environmental injustice, structural racism, and their implications on poverty, have already softened the space for the incoming of this virus to devastate people. You know, you think about something like water ... It should never have been turned off. It's one of the most frustrating things about the system of corporate capitalism ... You think about the logic of this –right?—and the realization that water should just be a human right for people, it should just be there for people, and then you fast-forward, and you think about the incoming pandemic, and we're telling people to wash their hands with warm, soapy water for 20 seconds. Well, if you don't have water in your house, you can't do that. All of those—all of that is seeded by decisions [structural-level factors] that have been made, that have been patterned around race and patterned around wealth for a very long time.

While the inequitable conditions associated with the racialized structure function to make many persons of color and other socially marginalized groups more vulnerable to COVID-19, police brutality, and killings of Black people, a persistent feature of the racialized US society, have continued unabated (Cohen, 2020; Waxman, 2020). One prominent example in this regard was the killing of George Floyd, who died on May 25, 2020, after a police officer kneeled on his neck for about nine minutes. Global protests erupted in response to Floyd's killing and to the racialized structures that make Black people more vulnerable to excessive policing, police brutality, and manifestations of structural racism. The intersection between this response to Floyd's killing and the disproportionate number of COVID-19 infections and deaths among Black people and other persons of color propelled the CDC to declare racism a public health crisis. At the same time, some scholars have noted that the nascent attention to the complex and enduring impact of structural racism also emphasizes limitations in our knowledge and efforts for addressing it (Cassels & van den Abbeele, 2021; Milano, 2020). Thus, there is an urgent need to develop and apply knowledge systems that respond to the complexities of structural racism if we are to fully understand and effectively confront it. To this end, we introduce a conceptual model – *structuragram* – as a heuristic to inform social work assessment and intervention and as a contribution to the *Grand Challenge to Eliminate Racism* (Teasley et al., 2021).

Structuragram: *a heuristic model to build structural competence*

Given the complexities of structural racism, we sought to develop a structural lens capable of considering the historical roots and economic foundations of racism (Piketty, 2020; Rosa & Díaz, 2019; Virdee, 2019).

Social work education has been misguided by "colorblindness, outdated conception of cultural competency, and white-centered history of liberal arts programs," with strategies intended to eradicate racism predominantly consisting of behavior-oriented interventions (Rosa & Díaz, 2019; Teasley et al., 2021, p. 15). To improve our competencies in addressing racism, we must recognize institutions and professional organizations as actors in reproducing racial inequalities independent of individuals' implicit biases due to institutionalized modes of meaning-making (Rosa & Díaz, 2019). Therefore, in addition to individuals and groups, institutions should also be a target of anti-racist education, practice, and research. We first consider the *History* of racism as critical to structural-level analysis to interrogate, interrupt, and eliminate it. Structurally, we metaphorically visualize racialized institutions as multiple conveyor belts. The directions and the destinations of these conveyor belts (e.g., healthcare system, education, housing, criminal justice system, media, the market) were established centuries before the current actors (Rosa & Díaz, 2019; Wilkerson, 2019b). Isabel Wilkerson (2019b) powerfully depicted US racialized structures as an old house with flaws in its original foundations. While we may not see all of the cracks, we must have the courage, as current occupants, to face what needs to be repaired. Structural racism or, as Wilkerson (2019b) put it, *caste*, is like the structure of a house, one built long before any of us were here. Far too many of us cannot see the ways in which the beams of the house manipulate race to facilitate differential life experiences and outcomes. It is not enough to simply obscure problems beneath a coat of paint – instead, we must assess the foundation. We must critically engage with the history of the structure to help us identify what constitutes the foundation, what has been inadequately patched, and what needs to be disrupted and reset. A critical element in engaging with the historical context of structural racism is discovering alternative paths to reimagine and reconfigure a transformative structure that will mitigate longstanding inequalities (e.g., poverty, hyper-incarceration, police brutality, and wealth inequities) and build a more socially just society. The *structuragram* moves beyond existing heuristics and tools, such as the ecomap and genogram, and the culturagram, which focuses on change within the individual (Congress, 2004; Hartman, 1995), and instead provides a systematic model for critically examining key structures related to communities, populations, phenomena, or events to identify interventions for structural change.

To be effective in applying the model (See, Figure 1), four primary questions should be explored: (1) What is the current community, population, phenomenon, or event to understand from a structural perspective? (2) How is race implicated in this story, particularly as it relates to the six domains presented? (3) How does the present connect to the past? Can you now see the structures? (4) What future can be imagined that would disrupt and dismantle the structures and processes that maintain structural racism? Once the current topic is identified, starting with the *Present* (Figure 1) event (e.g., COVID-19, police brutality, and health disparities), a meaningful connection with *History* should be established. This work is conducted within a context that fosters deliberative dialogue, not debate, to facilitate shared understanding and community and individual engagement with *History* and the *Present*. Such work is essential for raising critical consciousness and breaking the culture of silence that reinforces oppressive structures and practices (Ledwith, 2016). By identifying the patterns of inequities related to historical structures and processes,

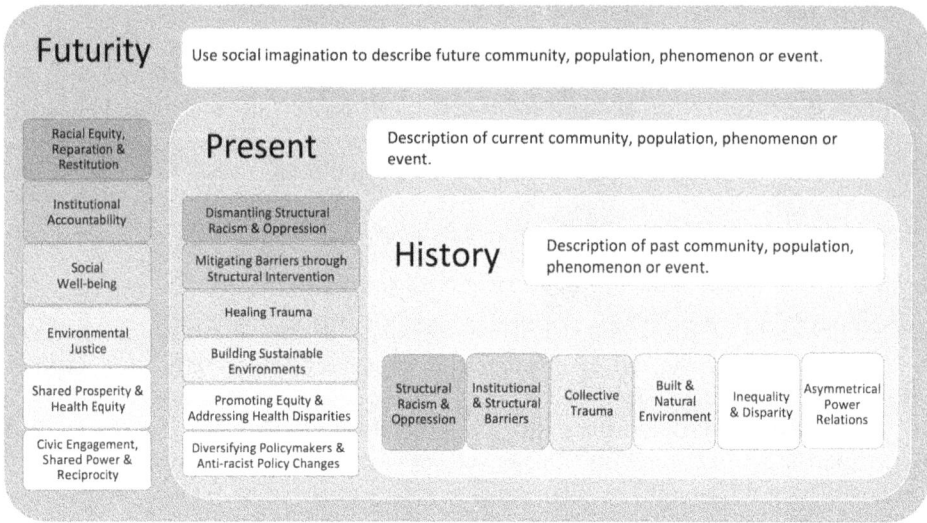

Figure 1. Structuragram.

participants can unpack the past and visualize the socially designed processes and structures serving as enabling factors. This will in turn, provide the context for mobilizing agency to elicit a deeper under-standing and motivate action. Participants will visualize the human design underlying the complexities of structural racism. Once the past and present structural patterns are made visible, participants can begin to map the *Future* by redesigning the structures and related networks based on the six domains (Finlev, 2012). As part of the critical action, opportunities are created to engage in bold steps to begin implementing the new designs while ensuring that the *Future* does not reproduce structural racism and the human suffering and injustice accompanying it. Various methods, such as storytelling, documentaries, heritage sites, and arts, can be used to complement this process to raise consciousness, generate a more profound under-standing, and prompt critical actions (See, Boddie, 2019; Hillier & Boddie, 2012).

In Figure 2, using the *Summer of Racial Reckoning*, we provide a case example of how the *structuragram* can be applied to structural-level assessment, analysis, and intervention to eliminate structural racism and its impact on communities, families, and individuals. We demonstrate the ways in which the *structuragram* can be used to practice structural competency.

Without the age of the cell phone and the sheltering-in-place mandated at the onset of the coronavirus pandemic, the murder of George Floyd in Minneapolis, Minnesota, would have remained invisible, and any response made silent (Present Event). However, as we look at the past, history repeats itself. The *Red Summer of 1919* was one of the most violent periods in US history, as riots in several cities from April to November 1919 left 250 Black people dead (Roberts, 2021). Like the 2020 *Summer of Racial Reckoning*, this earlier protest responded to the killing of a Black individual and unjust law enforcement in a Midwestern city. Teenager Eugene Williams was killed after he unknow-ingly crossed into a segregated area in Chicago, Illinois; his assailant, a white man, was not taken into custody by law enforcement. However, the state militia was called to contain the protesters. This tragic event reflects collective trauma as it took place on the heels of the 1918 Spanish Flu outbreak leaving 38 dead and over 1,000 homeless in Chicago. The panic that spread across the country was complicated by the health needs of a nation fighting for democracy. Although Black men served in the armed forces, they experienced limited freedoms upon their return home. The Jim Crow codes facilitated white-dominated law enforcement, racial exclusion, discrimination, and unequal access to health services, education, and housing. This gave way to segregated, often overcrowded neighborhoods and poor housing conditions with insufficient access to water for Black people in US cities. Social and economic disparities along with inadequate health access were the plight of Black people during 1918

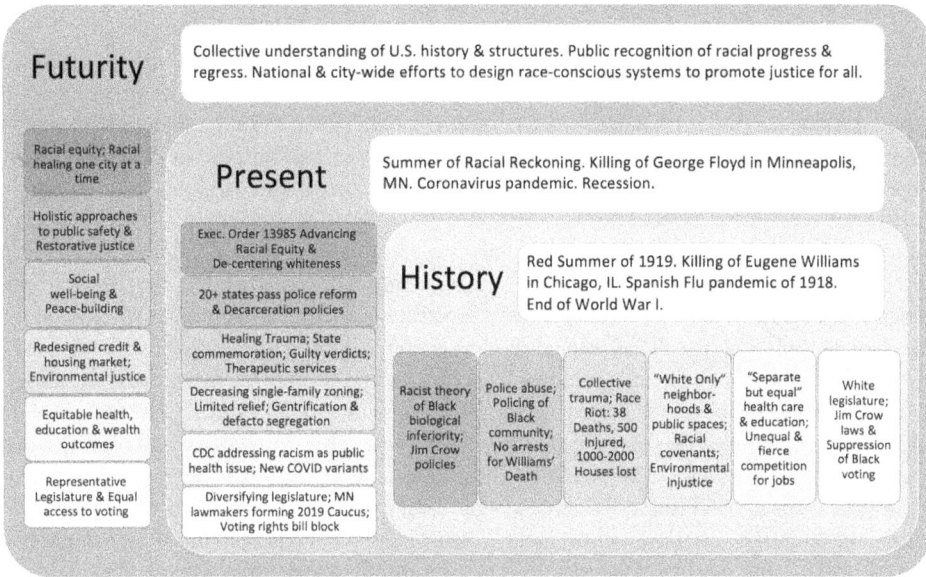

Futurity
Collective understanding of U.S. history & structures. Public recognition of racial progress & regress. National & city-wide efforts to design race-conscious systems to promote justice for all.

Racial equity; Racial healing one city at a time

Present
Summer of Racial Reckoning. Killing of George Floyd in Minneapolis, MN. Coronavirus pandemic. Recession.

Holistic approaches to public safety & Restorative justice

Exec. Order 13985 Advancing Racial Equity & De-centering whiteness

Social well-being & Peace-building

20+ states pass police reform & Decarceration policies

History
Red Summer of 1919. Killing of Eugene Williams in Chicago, IL. Spanish Flu pandemic of 1918. End of World War I.

Redesigned credit & housing market; Environmental justice

Healing Trauma; State commemoration; Guilty verdicts; Therapeutic services

Equitable health, education & wealth outcomes

Decreasing single-family zoning; Limited relief; Gentrification & defacto segregation

| Racist theory of Black biological inferiority; Jim Crow policies | Police abuse; Policing of Black community; No arrests for Williams' Death | Collective trauma; Race Riot: 38 Deaths, 500 Injured, 1000-2000 Houses lost | "White Only" neighbor-hoods & public spaces; Racial covenants; Environmental injustice | "Separate but equal" health care & education; Unequal & fierce competition for jobs | White legislature; Jim Crow laws & Suppression of Black voting |

Representative Legislature & Equal access to voting

CDC addressing racism as public health issue; New COVID variants

Diversifying legislature; MN lawmakers forming 2019 Caucus; Voting rights bill block

Figure 2. Example structuragram: 2020 summer of racial reckoning.

(Krishnan et al., 2020). Black people were at a higher risk for malnutrition and respiratory and pulmonary diseases. Public health officials justified the limited attention given to the health of Black people by concluding that the colored race was more immune to influenza than the White population. This same rumor circulated in the early days of the COVID-19 pandemic. During the 1919 Spanish Flu outbreak, Black people would often die faster, which was frequently ascribed to their inferiority (Krishnan et al., 2020). Blacks who lived in the Midwest at the time, including Illinois and Minnesota, were likely part of the silent revolution of the half-million Black people who left the south for northern cities to escape terror and discrimination.

Fast forward to 2007, where five Black police officers sued Minneapolis, alleging that city leadership tolerated discriminatory practices against Black people, including their own police officers. This case was settled out of court for over $800,000 (Alonso, 2020). Although the current police chief in Minneapolis was among these five officers, he has not been able to dismantle this system despite his efforts. This is because he is up against a network of racialized institutions (e.g., health care, criminal justice system, education) that must be engaged simultaneously, not independently. One year after George Floyd's death, former police officer Derek Chauvin was convicted of his murder and sentenced to 22.5 years in prison. As we connect the present more closely to the past, we note some changes occurring that appear to be intended to fix the cracks in several systems. President Biden signed Executive Order 13,985 Advancing Racial Equity and supported the George Floyd Justice in Policing Act of 2021. Over 20 states have passed police reform policies designed to restrict the use of force and/ or call for greater transparency and accountability. The John Lewis Voting Rights Act was also proposed while, at the same time, 20 new laws were introduced to restrict access to voting. As the pandemic persists, more deaths have occurred in 2021 than in 2020, and Black people and other people of color continue to be disproportionately impacted physically and economically due to unequal access to services and limited relief support. Several new proposals have been introduced to achieve a collective understanding of US history and structures and the public recognition of racial progress and regress to reimagine the future. A call for national and city-wide efforts to design race-conscious systems to promote justice for all is detailed using the six domains.

Future directions

The concepts of structural competency and the violence associated with structural racism underscore the urgency of developing additional language and new tools to build a post-COVID-19 future that works for the common good of all. This work is needed as social work curricula rarely prepare social work students "to think and intervene structurally" (Downey et al., 2019, p. 82). Structural competency calls for social workers and other healthcare professionals to identify the ways in which institutions, neighborhood conditions, city politics, market forces, public policies, and healthcare delivery systems shape both symptoms and diseases and to ultimately mobilize professionals to dismantle these inequalities, particularly beyond client–professional interactions (J. M. Metzl & Hansen, 2021). To accomplish this work will first require structural competency – the ability, skills, and willingness – to create authentic collaborations across and within various sectors as well as across racial, ethnic, class, and political divides. Taking this first step will allow social workers and other professionals, including clinicians and policymakers, to hear new stories about the nature of the problems and begin to change the narrative from vulnerability and scarcity to possibilities and abundance. The zero-sum game narrative no longer serves us. The *structuragram* allows us to see how structurally violent conditions perpetuate structural racism, ultimately affecting everyone. Instead of keeping our gaze focused on cultural differences, behavioral shortcomings, and biological predispositions, we now have the language and networks needed to make visible what has been invisible to some of us. We seek ways to visualize the structures at work in our communities, organizations, neighborhoods, various systems, cities, states, and the country as a whole. We must also become more invested in identifying opportunities in infrastructure to create robust structures for communities everywhere.

Building on the work of other scholars, we suggest the following ways to go upstream to begin to address the structures that perpetuate structural violence. First, a working group of community members and professionals across relevant healthcare sectors (schools, law enforcement, corrections, parks and recreation, housing, food system, urban planning, transportation) should be developed to create transdisciplinary collaborative and unified goals or engage in existing ones, like the Global Structural Competency Network (see https://structuralcompetency. org/). Second, stories (notably the stake in this work) from those participating across the network of community members and professionals should be shared to create a new narrative that offers new possibilities and assets. Third, data should be collected and visualized to document the geographies and populations affected by health inequities and to begin to imagine new possibilities. Furthermore, a more integrated community-based system of care should be established that includes the relevant sectors and local community members. Additionally, training should be provided to help members of collaborations learn to use public deliberation methods to discuss, craft, and influence policy. Indigenous and decolonizing research and practice methodologies should also be promoted to sustain equitable knowledge production. Of particular importance here is the need for social work education to help practitioners and researchers build structural competency skills that will enable them to assess problems and design interventions that recognize the complex and interactive relationships between the network of social institutions and organizations, as well as between individuals or communities. We must shift from efforts to control, simplify, or ignore complexity toward embracing it so that we can develop adaptive thinking and practices that will better position us to address pervasive and deeply embedded problems, such as structural racism. Social work educators must encourage the co-construction of new solutions by fostering inclusive classroom dialogs that allow diverse voices to be heard. Using the *structuragram*, social workers could provide the context for students and colleagues to take a historical inventory of social challenges and examine structural patterns visually by connecting the past and the present. Ultimately, by understanding the nature of the structural violence manifested as racism and the patterns deeply embedded in its foundation, solutions of an appropriate magnitude can be designed to achieve justice and structural change.

Racism comes at a high cost to everyone, as Heather McGhee (2021) reminded us. COVID-19 and the simultaneous police violence against people of color have demonstrated that risks and vulnerabilities to health and well-being are strongly linked to macro-level forces in racially stratified and complex ways. Therefore, unless we can begin to recognize and respond to the complexity of racism's structural violence as well as its collective and deleterious consequences, we will continue to lack the collective will and the competency to act. As a result, we will miss the opportunity presented by the current moment – the COVID-19 pandemic – to actualize the greatness expected of us as a profession and a discipline, and ultimately, as a nation.

Disclosure statement

No potential conflict of interest was reported by the author(s).

References

Alexander, M. (2020). *The new Jim Crow: Mass incarceration in the age of colorblindness (10th Anniversary)*. New Press.

Alonso, M. (2020, June 1). Minneapolis' top cops sued the department in 2007. Here's why it matters today. *CNN* . https://www.cnn.com/2020/06/01/us/minneapolis-police-chief-sued-department-2007-trnd/index.html

Bailey, Z. D., Feldman, J. M., & Bassett, M. T. (2021). How structural racism works — Racist policies as a root cause of U.S. racial health inequities. *New England Journal of Medicine, 384*(8), 768–773. https://doi.org/10.1056/nejmms2025396

Bailey, Z. D., Krieger, N., Agénor, M., Graves, J., Linos, N., & Bassett, M. T. (2017). Structural racism and health inequities in the USA: Evidence and interventions. *The Lancet, 389*(10077), 1453–1463. https://doi.org/10.1016/S0140-6736(17)30569-X

Bailey, Z. D., & Moon, J. R. (2020). Racism and the political economy of COVID-19: Will we continue to resurrect the past? *Journal of Health Politics, Policy and Law, 45*(6), 937–950. https://doi.org/10.1215/03616878-8641481

Bavel, J. J. V., Baicker, K., Boggio, P. S., Capraro, V., Cichocka, A., Cikara, M., Crockett, M. J., Crum, A. J., Douglas, K. M., Druckman, J. N., Drury, J., Dube, O., Ellemers, N., Finkel, E. J., Fowler, J. H., Gelfand, M., Han, S., Haslam, S. A., Jetten, J., & Willer, R. (2020). Using social and behavioural science to support COVID-19 pandemic response. *Nature Human Behaviour, 4*(5), 460–471. https://doi.org/10.1038/s41562-020-0884-z

Blankenship, M., & Reeves, R. V. (2020, July 10). From the George Floyd moment to a black lives matter movement, in tweets. *Brookings*. https://www.brookings.edu/blog/up-front/2020/07/10/from-the-george-floyd-moment-to-a-black-lives-matter-movement-in-tweets/

Boddie, S. C. (2019). *Unfinished Business: From the great migration to black lives matter*. Unfinished Business. https://www.unfinishedbusiness1916.com/

Booth, J. (2020). Structurally competent social work research: Considering research methods and approaches that account for a recursive relationship between individuals and structures. *Journal of Sociology and Social Welfare, 46*(4), 27–49. https://scholarworks.wmich.edu/jssw/vol46/iss4/4.

Braveman, P., Egerter, S., & Williams, D. R. (2011). The social determinants of health: Coming of age. *Annual Review of Public Health, 32*(1), 381–398. https://doi.org/10.1146/annurev-publhealth-031210-101218

Braveman, P., & Gottlieb, L. (2014). The social determinants of health: It's time to consider the causes of the causes. *Public Health Reports, 129*(Suppl 2), 19–31. https://doi.org/10.1177/00333549141291S206:

Calathes, W. (2017). Racial capitalism and punishment philosophy and practices: What really stands in the way of prison abolition. *Contemporary Justice Review: Issues in Criminal, Social, and Restorative Justice* 20(4), 442–455. https://doi.org/10.1080/10282580.2017.1383774

Cassels, S., & van den Abbeele, S. (2021). A call for epidemic modeling to examine historical and structural drivers of racial disparities in infectious disease. *Social Science and Medicine* 276, 113833 . https://doi.org/10.1016/j.socscimed.2021.113833

Centers for Disease Control and Prevention. (2020). *COVID-19 in racial and ethnic minority groups*. https://www.cdc.gov/coronavirus/2019-ncov/community/health-equity/racial-ethnic-disparities/disparities-illness.html

Chambers, J. E., & Ratliff, G. A. (2019). Structural competency in child welfare: Opportunities and applications for addressing disparities and stigma. *Journal of Sociology & Social Welfare, 46*(4), 51–75. https://scholarworks.wmich.edu/jssw/vol46/iss4/5.

Cohen, L. (2020, September 10). *Police in the U.S. killed 164 black people in the first 8 months of 2020*. These are their names. (Part I: January-April). CBS News. https://www.cbsnews.com/pictures/black-people-killed-by-police-in-the-u-s-in-2020/

Congress, E. (2004). Cultural and ethnic issues in working with culturally diverse patients and their families: Use of the *culturagram* to promote cultural competency in health care settings. *Social Work in Health Care, 39*(3/4), 249–262. https://doi.org/10.1300/J010v39n03_03

Czeisler, M. É., Lane, R. I., Petrosky, E., Wiley, J. F., Christensen, A., Njai, R., Weaver, M. D., Robbins, R., Facer-Childs, E. R., Barger, L. K., Czeisler, C. A., Howard, M. E., & Rajaratnam, S. M. W. (2020). Mental health, substance use, and suicidal ideation during the COVID-19 pandemic — United States. *MMWR. Morbidity and Mortality Weekly Report, 69*(32), 1049–1057. June 24–30, 2020. https://doi.org/10.15585/mmwr.mm6932a1

Deitz, S., & Meehan, K. (2019). Plumbing poverty: Mapping hot spots of racial and geographic inequality in U.S. household water insecurity. *Annals of the American Association of Geographers, 109*(4), 1092–1109. https://doi.org/10.1080/24694452.2018.1530587

Downey, M. M., Neff, J., & Dube, K. (2019). Don't just call the social worker: Training in structural competency to enhance collaboration between healthcare social work and medicine. *Journal of Sociology & Social Welfare, 46*(4), 77–95. https://scholarworks.wmich.edu/jssw/vol46/iss4/6

El-Sayed, A. (2020, March 31). Communities enduring racism & poverty will suffer most due to COVID-19 [Interview]. *Democracy now!* https://www.democracynow.org/2020/3/31/abul_el_sayed_epidemic_of_insecurity

Farmer, P. (1996). On suffering and structural violence: A view from below. *Daedalus, 125*(1), 261–283. http://www.jstor.org/stable/20027362

Finlev, T. (2012). Future peace: Breaking the cycle of violence through future thinking. *Journal of Futures Studies, 16*(3), 47–62. https://jfsdigital.org/wp-content/uploads/2013/10/163-A03.pdf

Fisher, K. (2021). An experiential model for cultivating humility and embodying antiracist action in a and outside the social work classroom. *Advances in Social Work, 21*(2/3), 690–701. https://doi.org/10.18060/24184

Fisher-Borne, M., Cain, J. M., & Martin, S. L. (2015). From mastery to accountability: Cultural humility as an alternative to cultural competence. *Social Work Education, 34*(2), 165–181. https://doi.org/10.1080/02615479.2014.977244

Galtung, J. (1969). Violence, peace, and peace research. *Journal of Peace Research, 6*(3), 167–191. https://doi.org/10.1177/002234336900600301

Garg, S., Kim, L., Whitaker, M., O'Halloran, A., Cummings, C., Holstein, R., Prill, M., Chai, S. J., Kirley, P. D., Alden, N. B., Kawasaki, B., Yousey-Hindes, K., Niccolai, L., Anderson, E. J., Openo, K. P., Weigel, A., Monroe, M. L., Ryan, P., Henderson, J., & Fry, A. (2020). Hospitalization rates and characteristics of patients hospitalized with laboratory-confirmed coronavirus disease 2019 — COVID-NET. *MMWR Morbidity and Mortality Weekly Report* 14 states. *69*(15), 458–464. March 1–30, 2020 http://dx.doi.org/10.15585/mmwr.mm6915e315

Gee, G. C., & Hicken, M. T. (2021). Structural racism: The rules and relations of inequity. *Ethnicity & Disease, 31*(Suppl), 293–300. https://doi.org/10.18865/ed.31.S1.293

Hartman, A. (1995). Diagrammatic assessment of family relationships. *Families in Society, 76*(2), 111–122. https://doi.org/10.1177/104438949507600207

Hillier, A., & Boddie, S. C. (2012). The Ward: Race and Class on Du Bois' Seventh Ward. *The Ward.* http://www.dubois-theward.org/

Jacobs, L. A., & Mark, H. (2019). Constructing the structurally competent classroom. *Journal of Sociology & Social Welfare, 46*(4), 125–145. https://scholarworks.wmich.edu/jssw/vol46/iss4/8.

Jacoby, S. F., Dong, B., Beard, J. H., Wiebe, D. J., & Morrison, C. N. (2018). The enduring impact of historical and structural racism on urban violence in Philadelphia. *Social Science and Medicine, 199*, 87–95. https://doi.org/10.1016/j.socscimed.2017.05.038

Jean-Baptiste, C. O., & Green, T. (2020). Commentary on COVID-19 and African Americans. The numbers are just a tip of a bigger Iceberg. *Social Sciences & Humanities Open, 2*(1), 100070. https://doi.org/10.1016/j.ssaho.2020.100070

Kohli, H. K., Huber, R., & Faul, A. C. (2010). Historical and theoretical development of culturally competent social work practice. *Journal of Teaching in Social Work, 30*(3), 252–271. https://doi.org/10.1080/08841233.2010.499091

Krishnan, L., Ogunwole, S. M., & Cooper, L. A. (2020). Historical insights on Coronavirus disease 2019 (COVID-19), the 1918 influenza pandemic, and racial disparities: Illuminating a path forward. *Annals of Internal Medicine, 173*(6), 474–481. https://doi.org/10.7326/M20-2223

Kyere, E. (2022). Racialized health care inequities dating to slavery. In R. D. Smith, S. Boddie, & B. English (Eds.), *Racializing health, COVID-19, and religious responses: Black Atlantic context and perspectives* (pp. 35–42). Routledge.

Laster Pirtle, W. N. (2020). Racial capitalism: A fundamental cause of novel coronavirus (COVID-19) pandemic inequities in the United States. *Health Education and Behavior, 47*(4), 504–508. https://doi.org/10.1177/1090198120922942

Laurie, E. W., & Shaw, I. G. R. (2018). Violent conditions: The injustices of being. *Political Geography, 65*, 8–16. https://doi.org/10.1016/j.polgeo.2018.03.005

Ledwith, M. (2016). *Community development in action: Putting freire into practice.* Policy Press.

Lopez, L., Hart, L. H., & Katz, M. H. (2021). Racial and ethnic health disparities related to COVID-19. *Journal of the American Medical Association, 325*(8), 719–720. https://doi.org/10.1001/jama.2020.26443

McGhee, H. (2021). *The sum of us: What racism costs everyone and how we can prosper together.* One World.

Medvedyuk, S., Govender, P., & Raphael, D. (2021). The reemergence of Engels' concept of social murder in response to growing social and health inequalities. *Social Science & Medicine, 289*, 114377. https://doi.org/10.1016/j.socscimed.2021.114377

Metzl, J. M., & Hansen, H. (2014). Structural competency: Theorizing a new medical engagement with stigma and inequality. *Social Science and Medicine* 103, 126–133. https://doi.org/10.1016/j.socscimed.2013.06.032

Milano, B. (2021, April 22). *With COVID spread, racism—not race—is the risk factor. Harvard Gazette.* https://news.harvard.edu/gazette/story/2021/04/with-covid-spread-racism-not-race-is-the-risk-factor/

Neff, J., Holmes, S. M., Knight, K. R., Strong, S., Thompson-Lastad, A., McGuinness, C., Duncan, L., Saxena, N., Harvey, M. J., Langford, A., Carey-Simms, K. L., Minahan, S. N., Satterwhite, S., Ruppel, C., Lee, S., Walkover, L., De Avila, J., Lewis, B., Matthews, J., & Nelson, N. (2020). Structural competency: Curriculum for medical students, residents, and interprofessional teams on the structural factors that produce health disparities. *MedEdPORTAL, 16* (10888), 1–10. https://doi.org/10.15766/mep_2374-8265.10888

O'Brien, R., Neman, T., Seltzer, N., Evans, L., & Venkataramani, A. (2020). Structural racism, economic opportunity and racial health disparities: Evidence from U.S. counties. *SSM - Population Health, 11*, 100564. https://doi.org/10.1016/j.ssmph.2020.100564

Piketty, T. (2020). *Capital and ideology.* Harvard University Press.

Reich, R. (2020a). *Covid-19 pandemic shines a light on a new kind of class divide and its inequalities.* The Guardian. https://www.theguardian.com/commentisfree/2020/apr/25/covid-19-pandemic-shines-a-light-on-anew-kind-of-class-divide-and-its-inequalities?utm_term=Autofeed&CMP=twt_gu&utm_medium&utm_source=Twitter#Echobox=1587878684

Reich, R. B. (2020b). *The system: Who rigged it, how we fix it.* Vintage.

Richardson, E. T., Malik, M. M., Darity, W. A., Mullen, A. K., Morse, M. E., Malik, M., Maybank, A., Bassett, M. T., Farmer, P. E., Worden, L., & Jones, J. H. (2021). Reparations for Black American descendants of persons enslaved in the U.S. and their potential impact on SARS-CoV-2 transmission. *Social Science and Medicine, 276*, 113741 . https://doi.org/10.1016/j.socscimed.2021.113741

Roberts, J.D. (2021, June 9). Pandemics and protests: America has experienced racism like this before. Brookings. https://www.brookings.edu/blog/how-we-rise/2021/06/09/pandemics-and-protests-america-has-experienced-racism-like-this-before/"

Robichaux, C., & Sauerland, J. (2021). Social determinants of health, COVID-19, and structural competence. *OJIN: The Online Journal of Issues in Nursing, 26*(2). https://doi.org/10.3912/OJIN.Vol26No02PPT67

Rosa, J., & Díaz, V. (2020). Raciontologies: Rethinking anthropological accounts of institutional racism and enactments of white supremacy in the United States. *American Anthropologist, 122*(1), 120–132. https://doi.org/10.1111/aman.13353

Rothstein, R. (2017). *The color of law: The forgotten history of how our government segregated America.* Liveright Publishing Corporation.

Sacks, T., & Jacobs, L. (2019). Introduction to the special issue on structural competency. *Journal of Sociology & Social Welfare, 46*(4), 1–4 https://scholarworks.wmich.edu/jssw/vol46/iss4/2.

Singer, M., & Rylko-Bauer, B. (2020). The syndemics and structural violence of the COVID pandemic: Anthropological insights on a crisis. *Open Anthropological Research, 1*(1), 7–32. https://doi.org/10.1515/opan-2020-0100

Smith Lee, J. R., & Robinson, M. A. (2019). "That's my number one fear in life. It's the police": Examining young Black men's exposures to trauma and loss resulting from police violence and police killings. *Journal of Black Psychology, 45* (3), 143–184. https://doi.org/10.1177/0095798419865152

Teasley, M. L., McCarter, S., Woo, B., Conner, L. R., Spencer, M. S., & Green, T. (2021). *Grand challenges for social work initiative: Eliminate racism* (Working Paper). American Academy of Social Work & Social Welfare. https://grandchallengesforsocialwork.org/wp-content/uploads/2021/05/Eliminate-Racism-Concept-Paper.pdf

Van Dorn, A., Cooney, R. E., & Sabin, M. L. (2020). COVID-19 exacerbating inequalities in the US. COVID-19 does not affect everyone equally. In the US, it is exposing inequities in the health system. *The Lancet Public Health, 395*(10232), 1243–1244. https://doi.org/10.1016/S0140-6736(20)30893-X.

Virdee, S. (2019). Racialized capitalism: An account of its contested origins and consolidation. *Sociological Review, 67*(1), 3–27. https://doi.org/10.1177/0038026118820293

Waxman, O. (2020, May 28) *George Floyd's death and the long history of racism in Minneapolis. Time.* https://time.com/5844030/george-floyd-minneapolis-history/

Wilkerson, I. (2020). *Caste: The origins of our discontents.* Random House.

Williams, D. R., Lawrence, J. A., & Davis, B. A. (2019a). Racism and health: Evidence and needed research. *Annual Review of Public Health, 40*(1), 105–125. https://doi.org/10.1146/annurev-publhealth-040218-043750

Williams, D. R., Lawrence, J. A., Davis, B. A., & Vu, C. (2019b). Understanding how discrimination can affect health. *Health Services Research, 54*, 1374–1388. https://doi.org/10.1111/1475-6773.13222

Wills, C. D. (2021). Addressing structural racism: An update from the APA. *Current Psychiatry, 20*(3), 43–46. https://doi.org/10.12788/cp.0110

World Health Organization (WHO). (2021). *Social determinants of health.* http://www.who.int/social_determinants/en/

From social justice to abolition: living up to social work's grand challenge of eliminating racism

Kristen Brock-Petroshius⬤, Dominique Mikell, Durrell Malik Washington Sr. and Kirk James

ABSTRACT

How can social work live up to the 13th Grand Challenge of Eliminating Racism? In this article we argue for the replacement of the predominant social justice paradigm with a framework for anti-racist social work praxis informed by abolitionist principles. The primary aim of anti-racist social work praxis needs to be the building of power in Black, Indigenous, or Brown and poor communities. We define additional praxis principles, including engaging with critical theories, advancing macro-approaches, targeting racism at the source, and developing interventions to eliminate and address the effects of racism. We end by sharing concrete anti-racist praxis tools.

Recently the American Academy of Social Work and Social Welfare adopted its 13th Grand Challenge: Eliminating Racism (Stevenson & Blakey, 2021). The Grand Challenges serve as visionary yet practical goals to guide all social welfare practice and research (Barth et al., 2019). They set social welfare apart as a discipline by emphasizing that we are not content simply building knowledge. Rather, we use scientific knowledge to change the conditions of society.

For many years, social work scholars and practitioners advocated that one of the grand challenges be the elimination of racism. The academy's initial response was that such a goal cut across all the grand challenges and could simply be incorporated throughout (Grand Challenges for Social Work, 2020). In light of the uprisings and renewed public conversations about racism that occurred following George Floyd and Breonna Taylor's murders in 2020, the Academy changed course and announced Eliminating Racism as its 13th Grand Challenge.

The adoption of this grand challenge also occurred during the COVID-19 pandemic, which has demonstrated the impact of racism on Black, Indigenous, or Brown and marginalized communities (Tai et al., 2021). Black and Latinx people have been hospitalized at 5-times the rate of white people (Reyes, 2020). Poverty and healthcare access are interconnected and significantly influence people's health and their quality of life––with those who live in more densely populated areas and households being at greater risk of virus transmission. The unemployment rate skyrocketed and small businesses serving low-income communities faced huge losses. Black, Indigenous, and Brown workers were also disproportionately represented among essential workers unable to social distance, thus being put at higher risk of virus exposure (Reyes, 2020). This global pandemic has once again made it clear how essential it is for Social Welfare to work toward the elimination of racism.

The adoption of this grand challenge is a step in the right direction and begs the question: How can social work live up to the grand challenge of eliminating racism? We begin by making explicit our conceptualization of racism and assess the limitations of social work's use of the social justice framework as a guide for anti-racist work. Next, we review the history and principles of abolition and explore how they can be applied to the new grand challenge. We argue that the central aim of anti-racist social work praxis needs to be the building of power in Black, Indigenous, or Brown and poor and/or marginalized communities. We offer specific principles as part of a broader framework of anti-racist social work praxis, each leading to further questions to guide and assess social work education, practice, and research. This paper then concludes with tools that can be utilized toward the actualization of anti-racist praxis.

Conceptualizing racism

Developing strategies to eliminate racism requires first understanding what racism is, how it functions, and what its effects are. Racial capitalism describes the interlocking global systems and practices that produce social and economic profit through the exploitation of people of color, especially Black and Indigenous people (Robinson, 2000). Through the theoretical framework of racial capitalism, we must understand racism materially, that it functions to produce wealth and poverty. We must understand racism ideologically, that power is maintained by making logics that serve ruling class interests into the "common sense" of the people. We must also understand racism historically, that contemporary social forces are extensions of historical processes of slavery, settler colonialism, and colonization (Pulido, 2017). Racial capitalism as a framework puts economic, political, and ideological power at the center of understanding structural racism.

Structural racism consists of "the macro level systems, social forces, institutions, ideologies, and processes that interact with one another to generate and reinforce inequities among racial and ethnic groups" (Gee & Ford, 2011, p. 116; powell, 2008). In the United States, structural racism is continuously reproduced through group-differentiated relationships to poverty, displacement and dispossession, policing and incarceration, and immigration policy; segregation in housing, education, employment, and healthcare; voter disenfranchisement; vulnerability to community and state violence; and exposure to environmental pollutants (Omi & Winant, 2015). It is through systems, social forces, institutions, and ideologies that race itself is constructed.

These frameworks of racial capitalism and structural racism make clear that those targeted by racism are Black, Indigenous, or Brown communities that are poor and/or marginalized based on ethnicity, religion, sexuality, gender, ability, age, nationality, or geography. Structures of oppression are interlocking and anti-oppressive practice must be intersectional. Throughout this paper we refer specifically to Black, Indigenous, or Brown communities that are poor and/or marginalized and use the terms interchangeably.

Racial formation––the sociohistorical process through which racial categories are formed––is not static across time or place (Omi & Winant, 2015). Rather, the institutions and ideologies that structure racism and construct race itself are continuously adapted by political actors in response to macro-level forces such as changing laws and social norms, emerging technologies, and social movements. This process of racial formation involves a pattern of conflict and accommodation, repression and incorporation between social movements and the state. Over time, these processes lead to a rearticulation of ideology and changes to institutions to appease movements and dilute their power, altering the particular mechanisms that structure racism and construct race. Changes following the Civil Rights movement––from overtly racist to colorblind ideology and from explicit racial segregation to nondiscrimination policies that require proof of racist intent–– exemplify the patterns described in racial formation theory (Omi & Winant, 2015). Institutions are continuously being remade.

The institutions that structure racism––schools, prisons, social service agencies, and banks, for example––are heavily shaped by policies. The amount of funding a school has, whether there are mental health professionals or police on staff, or whether Ethnic Studies is taught, for example, are decisions all dictated by policy. In moving away from framing racism solely as prejudice or racial animus, individual racism can be understood to include support for structurally racist policies (Kendi, 2019; Lipsitz, 2006).

Colorblind ideology, which downplays the salience of racism and focuses on having "good intentions," is often used to justify an individual's support for structurally racist policies (Bonilla-Silva, 2010). At the level of individuals, social work can be examined through this understanding of racism–– support for structurally racist policies that is often bolstered by colorblind ideology.

Hegemonic constructions of social justice and anti-racism

A social justice paradigm often encompasses social work's efforts to address racism. As one of our core professional values, social justice broadly asserts that we need to challenge injustice and pursue social change (National Association of Social Workers [NASW], 2017). Social justice frameworks are often enmeshed with or co-opted by liberalism, a framing of justice that focuses on hopes and aspirations while lacking critical and structural analysis of problems (Roy, 2006). Within social work, the definition of injustice, the methods for challenging it, and the type of social change we are pursuing are often left ambiguous. Social justice frameworks can be used for a variety of contradictory purposes: to call for transformation or to justify the status quo, to promote collective liberation or push for individual responsibility, and to center a top-down advocacy approach or a bottom-up organizing approach to achieving justice (Reisch, 2014). Nearly anything a typical social worker does could be framed as promoting social justice, particularly if working with communities of color. Much of social work practice, however, actively strengthens structural racism or passively reinforces it through lack of anti-racist action (Miller & Garran, 2017).

The history of social work is rooted in active racism. White social workers operated the boarding schools and child welfare system responsible for the forced removal and assimilation of Indigenous children from their families and nations. The United Nations defines these colonial practices as elements of genocide, resulting in lost languages, cultural traditions, and religious practices of entire generations and nations (Thibeault & Spencer, 2019). Black settlement house workers and the National Association of Black Social Workers (NABSW) have long challenged the dominant ideologies of the social work profession, advocating for the inclusion of communities of color in the settlement house agenda, more Black social workers on staff and in positions of power, support for Black Power movements, and divestment from oppressive systems (Bell, 2014). There are numerous examples of social work's active racism through the present day––including our role in operating Japanese Internment Camps, prisons and jails, and migrant family detention centers. White people continue to dominate most positions of power within the social work profession, even under the contemporary social justice paradigm.

The social justice framework is often hegemonic. Hegemony is the maintenance of power by ruling people (white elites in the United States) by turning the dominant ideology that supports structural oppression into the "common sense" of the public (Gramsci, 1971). The beliefs, perceptions, values, morals, and explanations of those in power are taken on by large numbers of the broader population despite occupying a very different material reality from those in power. In the context of social work, social justice becomes hegemonic when carceral logics are extended in service provision, when diversity or work with people of color satisfy the optics of anti-racism without requiring deeper interrogation of the work's substance, and when we lack the imagination and practices to actually work toward the elimination of racism.

Social work has defined anti-racist practice with similar ambiguity as social justice, leaving it vulnerable to becoming hegemonic through liberalism. Until recently the Educational Policy and Accreditation Standards (EPAS), the standards that guide all social work education, lacked any mention of racism or anti-racist practice altogether (Council on Social Work Education [CSWE], 2015).[1] In a report from the National Association of Social Work (NASW) specifically on action to address institutional racism, anti-racist action was described mostly on the levels of personal awareness, education, and organizational change. The one mention of addressing broader structural racism simply described "promote change" and "use the available resources to challenge racist policies, practices and behaviors" (Craig de Silva & Clark, 2007, p. 21). What is absent speaks volumes: there is no theory of change, no suggestion of tactics such as community organizing, no naming of goals such as building power in Black, Indigenous, or Brown and poor communities, and no defining of targets such as white communities or political actors. Specificity about anti-racist social work practice is critical; without it, hegemonic notions of racism and anti-racist practice will flourish and social work will fall prey to bolstering structural racism rather than eliminating it (Richie & Martensen, 2020). Abolition, because of its specificity in theory, goals, and practices, offers a stronger conceptualization of anti-racist social work practice.

Abolition praxis

Abolition is a vision and organizing practice intellectually rooted in the multiracial, Black-led Abolitionist Movement of the 18th and 19th centuries that overthrew the institution of slavery. The contemporary abolitionist movement is an extension of this radical Black tradition and seeks to eliminate the use of policing, imprisonment, punishment, and surveillance as institutions and practices that oppress Black, Indigenous, Brown and marginalized communities (Kelley, 2002). Historical and contemporary abolitionist movements understand that full abolition requires a dismantling of racial capitalism––interlocking global systems and practices that produce social and economic profit through the exploitation of people of color, especially Black and Indigenous people (Robinson, 2000; Sinha, 2017).

Both W.E.B. DuBois and Angela Davis studied that particular period of the Abolitionist Movement to further refine a conception of abolition. As Davis asserts, the primary project of abolition is not "a negative process of tearing down" but rather one of collectively "re-imagining institutions, ideas, and strategies, and creating new institutions . . . that render prisons obsolete" (Davis, 2005, p. 75). Neither just theory nor practice, abolition is often discussed as praxis––the ongoing process of using critical theory to inform strategy and practice, then reflecting on that work to further refine theory. Abolition implores us not only to dismantle oppressive institutions but also to build new ways to prevent and respond to harm, create justice, and build truly democratic political power (Du Bois, 1935; Davis, 2005). It is through this realization of self-determination, this building of political and economic power in Black, Indigenous, or Brown and poor communities, that abolition-democracy takes root.

While academic and organizing literatures on abolition consist of different perspectives and nuances, abolitionists converge on three core tenets. First, abolition takes seriously the power of and damage caused by the carceral state. Carcerality is understood to be about social regulation and the maintenance of political power more than it is about addressing crime or enhancing public safety. The expansion of carceral institutions is directly related to divestment from community resources that would enhance health, expand opportunities, and create equity (Richie & Martensen, 2020). The carceral state is understood as an extension of settler colonialism and slavery, oppressing Black, Indigenous, or Brown and marginalized communities (Hernandez, 2017). It is upheld through dominant ideology, the ways our hearts and minds are saturated with carceral logics of control, punishment, and disposal (Kaba, 2021).

A second tenet of abolition is the need to be vigilant about the ways in which advocated changes to carceral institutions can indirectly rely on punishment, sometimes leading to even further expansion or entrenchment with the legal punishment system. Reform efforts can easily be co-opted by the state, even parroting abolitionist language in referencing restorative justice, for example. Rigorous assessment of the impact of any attempted changes to carceral systems is essential, regardless of the stated intention. Approaches to address the host of social issues communities deal with must be developed in ways that do not fall prey to logics of punishment and control (Critical Resistance, 2020; Jacobs et al., 2021; Richie & Martensen, 2020).

A third tenet of abolition is the importance of centering those directly targeted by the carceral state. Organizing centers Black, Indigenous, or Brown and poor people who have been incarcerated, their families, and their broader neighborhoods and communities. The practice of abolition includes community self-determination, collective decision-making, and the building of power in communities targeted by the carceral state (Richie & Martensen, 2020).

In the wake of George Floyd's murder by Minneapolis Police, over 1,700 protests occurred in all 50 U.S. states and 40 countries – pushing the world to imagine abolitionist futures (Smith et al., 2020). The defunding, disbanding, or abolition of law enforcement is being publicly wrestled with in ways previously unimagined. The Minneapolis City Council voted to disband their police department (The Associated Press, 2020). San Francisco's mayor called for all non-emergency 911 calls to be dispatched to non-police officers (Dolan, 2020). New York City proposed a $1 billion cut to their police department, for a reduction of 17% (Rubinstein & Mays, 2020). Los Angeles County voters passed a ballot initiative that will divert 10% of unrestricted funds away from law enforcement and into community services and programs (Cosgrove, 2020). These public conversations and policy debates bring to light several core tensions between abolition praxis and a more liberal version of the social justice paradigm.

Both abolition and liberal social justice paradigms acknowledge the problems of racism and the carceral state, are outraged by police murders (at least of people without a weapon), and want to see changes to carceral systems. When detailed solutions are proposed, however, the distinctions between liberal social justice and abolition frameworks become more pronounced. Abolitionist policies are characterized as those that reduce the funding and scale of carceral institutions, challenge the notion that policing, surveillance, and incarceration increase safety, and reduce the tools, tactics, and technology available to carceral institutions (Critical Resistance, 2020). In this framework, which is widely used by abolition organizers, community policing, body cameras, and more training are "reformist reforms," while reducing the size of the police force, increasing funding for community health, education, housing, and services, and withdrawing police participation in militarization programs are abolitionist policies, even if their impacts are incremental. Abolition requires specificity in assessing policy impacts, preventing social justice frames from bolstering further expansion of carceral state expansion.

These tensions between abolitionist and "reformist reform" social change efforts became clear within social work during the summer 2020 Black Lives Matter uprisings. NASW put out a statement calling for improved training of the police and greater collaboration between social workers and law enforcement. Many social workers were outraged by this hegemonic stance and publicly critiqued the association for suggesting changes that would ultimately strengthen carceral institutions (Dettlaff & Abrams, 2020; Rasmussen & James, 2020). NASW saw themselves as promoting social justice through their approach of becoming more entrenched with law enforcement, demonstrating a clear failure of the social justice paradigm to necessitate anti-racism.

Anti-racist social work praxis

In contrast to the social justice paradigm, abolition would require that social welfare examine and reconceptualize our strategies, theories, practice methods, and research approaches with much greater specificity toward eliminating structural racism. Abolition calls on us to situate our work in its broader historical context and to engage in praxis. It also requires us to take seriously the power and damage caused by white supremacy, to be vigilant about the ways well-intentioned change efforts can reinforce structural racism, and to center those targeted by white supremacy. These core tenets lead to a vision for anti-racist social work praxis that builds power in Black, Indigenous, or Brown and marginalized communities––the embodiment of abolition democracy and social work's value of self-determination. Additional principles further support this primary aim and include the following: 1) engage critical theories to inform education, practice, and research 2) advance macro-level approaches of organizing, advocacy, and social movement mobilization, 3) target racism in dominant communities and institutions, and 4) develop interventions to eliminate and address the effects of racism. Our conceptual model is shown in Figure 1.

Build power in black, indigenous, or brown and poor communities

The relationship between social work and Black, Indigenous, or Brown and poor communities lies within a context of disfigured relationships chiefly created by chattel slavery and settler colonialism in the United States (Wilkerson, 2020). These disfigured relationships have caused white people to be

Anti-Racist Social Work Praxis

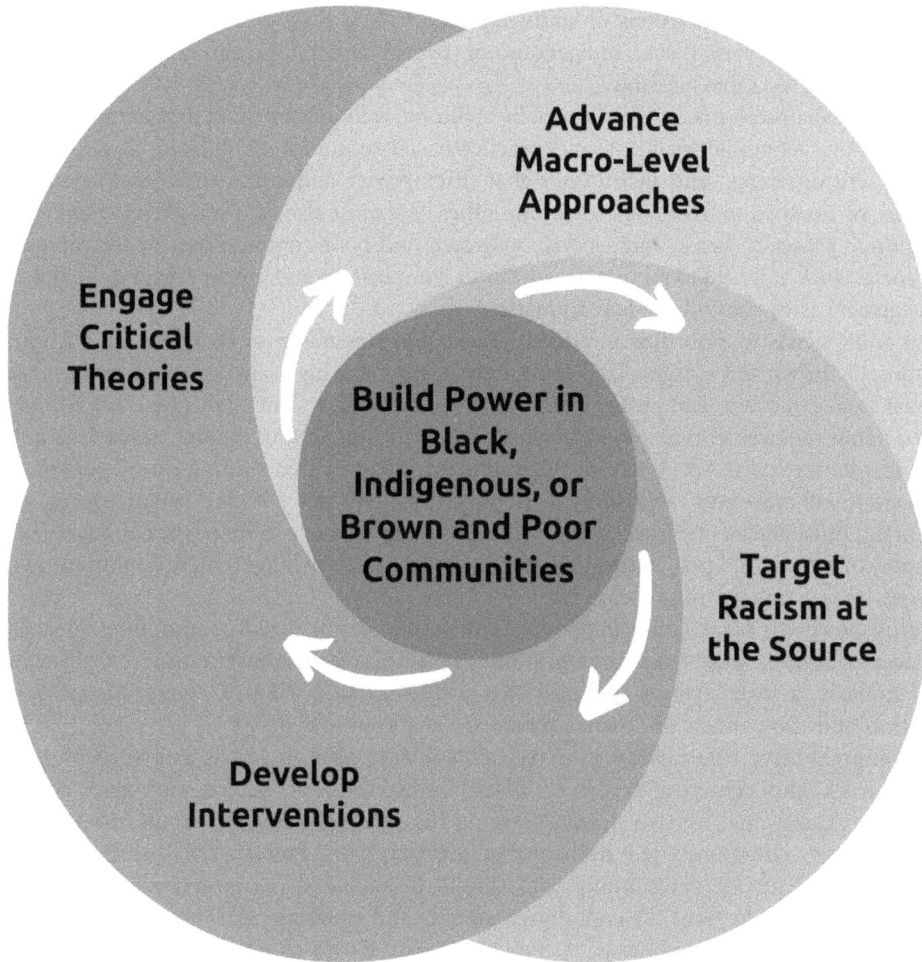

Figure 1. Anti-racist social work praxis is a non-linear process with a central aim of building power in Black, Indigenous, or Brown and poor communities. Additional, interconnected principles are engaging critical theories to inform education, practice, and research; advancing macro-level approaches of organizing, advocacy, and social movement mobilization; targeting racism in dominant communities and institutions; and developing interventions to eliminate and address the effects of racism.

"accustomed to the unearned deference from the subjugated group" (Wilkerson, 2020, p. 51). All other groups in the United States are pushed to survive by aligning their interests with the subjugator and distancing themselves from the most subjugated. Social workers as a profession often operate in the buffer zone of these relations, enacting the agenda of white elites by providing services that manage and prevent rebellion of the subjugated (Kivel, 2017). Social work operates within, perpetuates, and benefits from a social value system that grants those who are white and hold resources the right to make decisions on behalf of all other communities. This results in what Freire refers to as false charity, generosity that is nourished by and therefore can never be the solution to injustice, death, despair, and poverty (Freire, 1970).

The social work profession must work to dismantle disfigured relationships, false charity models, and social justice practices that continually undercut power-building efforts within Black, Indigenous or Brown and marginalized communities. Power, defined in its most simple form, is the ability of a community to make decisions about the things that affect them (Magill &

Clark, 1975). Centering communities directly targeted by structural racism--which requires social work to take seriously issues of grassroots political power--helps ensure that our efforts do not become well-intentioned extensions of white supremacy. It is through this central aim of building decision-making power in Black, Indigenous, or Brown and poor communities that anti-racist social work praxis becomes possible.

Anti-racist social work praxis, as informed by abolition, calls on us to both dismantle and build. We must use all the economic, political, and practice-based resources we hold to dismantle policies, practices, cultural norms, and institutions that disempower and commit violence against Black, Indigenous, or Brown and marginalized communities. We must also support efforts to build political and economic power in Black, Indigenous, or Brown and poor communities. These commitments require social work as a field to rebuild its workforce composition and pipeline, agency level practices, and engagement with community-based practices.

Social work needs to prioritize the development and leadership of social workers from Black, Indigenous, or Brown and marginalized communities. As a starting point, this requires reconsidering how social work education and professionalization are financed and implemented, including addressing curriculum requirements and credentialing processes. This shift in the workforce does not mean there is no role for social workers from the dominant group or that shifting social workers' demographics alone will make anti-racist social work praxis possible. Many of the normative forms of social work practice must end. Social workers can no longer hold the power to shape the conditions of Black, Indigenous, or Brown and poor communities. This power must belong to these communities even if our jobs as we know them must end.

Building power in Black, Indigenous, or Brown and marginalized communities means changing power structures within our organizations. This requires not just transparency but organizational honesty, a high level of openness that includes the sharing of organizational operating information and the context of how resources are acquired and decisions are made (Carruthers, 2018). Comprehensive information must be accessible to all individuals impacted by the organization's work and centering access to Black, Indigenous, or Brown and poor community members. Organizational honesty would serve as the starting point for experiments in creating truly democratic, community-led decision-making practices. Practically, this is no easy task. There is no blueprint for determining who should be included in community-led processes or what practices should be used through disagreements and conflicts. As Miriame Kaba encourages us, "We need a million experiments, A bunch will fail. That's good because we'll have learned a lot that we can apply to the next ones" (Hooks, 2020). Abolition praxis requires evolving reflection and action with the communities we aim to work with toward liberation--prioritizing dialogue, reflection, and communication and ultimately trusting Black, Indigenous, or Brown and marginalized communities and their abilities to reason and make decisions.

Building power includes the creation of new institutions and practices for community healing, accountability, and resource-sharing. Indigenous, Black, and Brown communities and abolitionist healers and organizers have already been leading the way. Healing justice efforts aim to holistically address the consequences of oppression and trauma on our bodies, minds, and spirit, supporting us in dismantling the causes of these harms without recreating them with each other (Carruthers, 2018). Transformative justice involves collective, community-based accountability and responses to harm that center healing for survivors and behavioral change for those that have done harm (Kim, 2019). Mutual aid is social movement work that includes the building of community networks to ensure people have the resources needed to survive--including reappropriating vacant, publicly owned homes for housing or organizing medical care and disaster relief (Spade, 2020). These practices can build power in Black, Indigenous, or Brown and poor communities through the collective creation of solutions to social problems that don't rely on dominant institutions.

Engage critical theories to inform education, practice, and research

Scientific theories from psychology and public health are commonly used in social work. While useful in developing individual-level interventions, they are often devoid of a more critical understanding of the broader historical context and social processes in which social problems are situated. Critical theories addressing race, racism, and anti-racism provoke a deeper wrestling with questions necessary for social work practice to be more effective (Constance-Huggins, 2012). These theories have largely been developed in ethnic and social science disciplines and include but are not exclusive to Critical Race Theory. Some of the theories discussed in this paper, for example, – racial capitalism, hegemony, racial formation, and abolition – are critical theories about racism but are not part of the specific legal and educational canon of Critical Race Theory, per se. Anti-racist social work praxis necessitates a deeper engagement with critical theories to inform our understanding of social problems and assist communities in developing solutions.

Critical theory, if applied as part of anti-racist praxis, would be particularly helpful in safeguarding against social work education, practice, and research directly or indirectly reinforcing structural racism. In the classroom, curricula would be rooted in understanding how systems have historically and currently sanction the violence, genocide, erasure, deprivation, and dispossession of Black and Indigenous peoples' lands, resources, and histories (Abdulle & Obeyesekere, 2017). Learning approaches would center the liberatory visions and learning desires of Black, Indigenous, or Brown and marginalized communities. In practice, the underlying ideologies and logics that inform intervention aims and theories of change would be made explicit. Many interventions would be assessed as harmful or dangerous, such as the separation of children from their families through the child welfare system. Interventions would address aims that have historically been marginalized in social work, such as the building of political and economic power in Black, Indigenous, or Brown and poor communities. In research, it would be required that inquiry aim to change structural conditions, not just study problems, and that studies be conducted through accountable community collaboration. Scholars would be questioned about their connections and commitment to communities they study.

Advance macro-level approaches

Direct services are an essential component of social work and "service is not liberation" (Richie & Martensen, 2020, p. 15). Any individual-level work must also include a macro-level component that aims to eliminate structural racism through policy change. These policies directly impact the individuals, communities, and organizations we serve and can either further enhance structural racism or work toward its elimination. Macro-level interventions such as community organizing, policy advocacy, and social movement participation must become central to the field of social work. Social work can play a role in advancing abolitionist policy agendas––including reducing the power and funding of carceral institutions, expanding voting rights, and transferring wealth back to individuals and communities who have been dispossessed, displaced, and exploited. By engaging in organizing and social movements, social workers can serve as part of a social force much broader than any discipline or institution.

As discussed earlier, we must address power dynamics by engaging and sharing power with those directly impacted by the particular policies we are aiming to address. For example, organizing and advocacy to weaken carceral institutions needs to center the leadership of and be accountable to Black, Indigenous, or Brown and marginalized people and families who have been or are under control of the carceral state. Efforts to abolish the child welfare system need to center the leadership of and be accountable to Black, Indigenous, or Brown and poor youth and families who have been or are under its control. The goal of any macro-level approach needs to be understood as radical policy and culture change to undermine structural racism, not simply liberal reforms that result in softer forms of racism.

Target racism at the source

Eliminating racism requires not only building power in Black, Indigenous, or Brown and poor communities, but also targeting racism at its source--white communities, political actors, and others who support status quo structural racism and dominant institutions. Without this approach, many social work interventions--if focused solely on the targets of structural racism--are simply developing band-aid solutions. Civil Rights and Black Liberation movement leaders have recognized this need as well, calling on white people serious about ending white supremacy to organize white communities for racial justice (American Radio Works, 1966). We have to move beyond people and also explicitly target structurally racist policies. White people must do the individual and collective work to dismantle white supremacy and hold other white people accountable to do the same. More broadly, those of us who are not the targets of anti-Blackness, settler colonialism, xenophobia, Islamophobia, anti-Semitism, colorism, classism, ableism, homophobia, and transphobia need to organize to end these forms of oppression in our own communities. By owning white supremacy as a problem of dominant communities, social work can more strategically attack racism at the source.

To attack racism at its source, we need skilled, accountable leaders willing to take risks to make systematic changes--not complacent people in power doing the bare minimum and hiding behind performative initiatives like most trainings on Diversity, Equity, and Inclusion. Social work can play a role in training and developing leaders equipped to tackle structural racism, which will then have impact both within and beyond our field. Leaders must target structurally racist, caste-like systems that keep white people at the top and people of color at the bottom. Organizing, critical theory, and internal work are essential for white people to explicitly target structurally racist policies, practices, and ideologies at the level of broader communities.

Develop interventions to eliminate and address the effects of racism

Anti-racist praxis informed by abolition also pushes us further in our discipline's unique goal within the social sciences to transform social structures, not just understand how they work. To this end, our field needs to become focused on developing, implementing, and evaluating interventions to weaken and ultimately eliminate racism. Senator Elizabeth Warren called for such an approach by stating, "It is time we start treating structural racism like we would treat any other public health problem or disease: investing in research into its symptoms and causes and finding ways to mitigate its effects" (Major, 2020). NASW recently called on the Biden-Harris administration to create a task force and establish government funding to address racism as a public health crisis (2021). Social work can apply the scientific rigor we use to address other social problems to the grand challenge of eliminating racism.

Much of social work efforts to address racism focus on understanding and mitigating the effects of racism on health outcomes in communities of color. Improving community health is an important intervention to address racism, but it cannot be the only approach. Additional interventions need to focus on building political and economic power in Black, Indigenous, or Brown and marginalized communities. Social work can play a role in voter mobilization, community organizing, and social movements and in the practices of collective decision-making, healing justice, and transformative justice. Social work can support the building of community wealth through the expansion of affordable housing, government-funded healthcare, education, and caregiving programs; attempting new economic forms outside the domains of capitalism such as mutual aid, cooperative economics, and guaranteed minimum incomes; and redressing historical harms through reparations of money and land to Black, Indigenous, or Brown and marginalized communities.

As described earlier, any efforts to address the effects of racism without focusing on the source of the problem function as band-aids. Social work also needs to develop interventions focused on addressing racism in white communities. Anti-racist interventions could include organizing strategies to mobilize white people into action as part of broader multiracial movements and to increase support

Table 1. Guiding questions for anti-racist social work praxis.

Central aim: Build power in Black, Indigenous, or Brown and poor communities
- How does this praxis dismantle policies, practices, systems, or structures that restrict the power of Black, Indigenous, or Brown and marginalized communities? How does this work strengthen power and accountability within and to communities?
- Are Black, Indigenous, or Brown and poor community members represented at all decision-making levels, including setting goals, managing resources, and assessing progress?
- Are truly collective, democratic decision-making practices being strengthened through this praxis?

Principle 1. Engage critical theories to inform education, practice, and research
- What is the racialized impact of this praxis, regardless of intention?
- What logic and theory of change does this praxis suggest, even implicitly?
- How can we anticipate attempts by dominant institutions to co-opt, repress, or contain this work?

Principle 2. Advance macro-approaches of organizing, advocacy, and movement mobilization
- How does this praxis engage micro and macro-level social workers to advance systems change through policy?
- How do we center the individuals, families, and communities directly impacted by the systems we seek to change in our organizing and advocacy?
- How do we share power, resources, and compensate community stakeholders for their engagement?

Principle 3. Target racism at the source
- How does this praxis move us from just imagining a world that is anti-racist to taking action to achieve this goal?
- What strategies, interventions, and research does this praxis develop to eliminate racism in white communities and dominant institutions?
- What does accountability to Black, Indigenous, or Brown and marginalized communities look like in this praxis?

Principle 4. Develop interventions to eliminate and address the effects of racism
- How does this praxis develop interventions to strengthen political power, economic resources, and health outcomes in Black, Indigenous, or Brown and poor communities? Who is making these assessments and decisions?
- How does this praxis develop interventions to move white people and other dominant group members to stop racist behaviors, move from passive support for the status quo into anti-racist action, and increase support for racial equity policies and power-building in Black, Indigenous, or Brown and marginalized communities?
- Does this praxis reflect bold imagining and not simply the type of work that is funded or valued by dominant institutions?

for racial equity policies. Anti-racist interventions could also take the form of therapeutic work to develop individual and collective competencies for stopping racist behaviors and engaging in anti-racist action as well as organizational development work to enhance anti-racist structures, policies, and practices. Additional interventions could focus on addressing anti-Blackness, settler colonialism, xenophobia, Islamophobia, anti-Semitism, colorism, classism, ableism, homophobia, and transphobia in communities not targeted by these specific forms of oppression.

Taken together, the central aim and further principles of anti-racist social praxis provoke further questions. These guiding questions can be used to assess the extent to which our praxis--the ongoing process of action and reflection--furthers the elimination of structural racism. These guiding questions are found in Table 1.

Tools for anti-racist praxis

As social workers, we know anti-racist work is challenging! However, we unequivocally believe that this work must happen for any of our work to be effective. Social work's grand challenge of eliminating racism has implications for all of us regardless of racial identification and economic stratification. *Humanity* has been divided, subjugated, and controlled by racism. Furthermore, our collective future--or it's possibility, very much depends on our willingness to individually and collectively acknowledge, heal, and dismantle the vestiges of racism in our ideologies, institutions, interpersonal ways of being, and internalized self. Here we share a praxis framework centered around bias, intention setting, and ground rules (B.I.G.) that has been beneficial to us on our journey to eliminate racism and other vestiges of oppression from within our practice.

Social work often posits that we must see people (including ourselves) within their environment--yet what does that mean when discussing and actualizing the work to become anti-racist practitioners? Activist Adrienne Maree Brown similarly ponders that question and concludes that the systems we seek to transform are "peopled, in part, by the same flawed complex individuals that

I [we] love" (Brown, 2020, p. 68). In her acknowledgment and reflection, brown discusses the necessity of humility in how anti-racist practitioners see themselves, each other, and the tools of transformation.

In understanding that we all have unlearning to do, we must critically examine our formal and informal educational systems for conscious and unconscious bias. Education can be utilized to maintain the status quo or as a tool of liberation (Freire, 1970). Most of us are taught to absorb knowledge without critical engagement, the banking system of education that maintains the ideologies and practices of oppression. The first tool we can utilize in anti-racist work is thus a recognition that we all come to it with various lived experiences often convoluted with some degree of Bias––whether racial, gender, sexual, cultural, or political.

It is through recognition and action toward dismantling bias (including our own) that we can show up with the humility necessary to engage in liberatory practices. bell hooks cautions that this unpacking is not going to be easy and asks that we commit to "brave space" (hooks, 1994). Brave space is a process that must be operationalized and co-created by participants and involve as a primary component Intention setting. This can be something as simple as asking each member to declare their intentions or values before engaging in anti-racist work. Clearly expressed and co-created intentions allow us to create *accountability* toward brave space.

Recognizing *bias* and *intention* setting are precursory steps to creating brave space, given that it is not a matter of "if" but "when" conflict will occur (hooks, 1994). The final tool we will share is called Ground rules––a living document co-created by practitioners committed to anti-racist praxis. Ground rules outline how committed anti-racist practitioners will engage as we collectively seek to understand varied lived experiences and responses to harm while taking steps toward collective healing and liberation from oppressive systems. Examples of co-created ground rules are as follows: 1) Committing to call in vs. call out, 2) Using human-centered language, 3) Seeking understanding, 4) Committing to critical, reflective, and vulnerable communication, and 5) Honoring our collective healing process.

The last ground rule is of particular importance for social workers seeking to eliminate racism––a sickness weaved into the very fabric of the United States. Tackling racism at the root will trigger, provoke, and create immense tension within our communities. However, tension is not a bad thing and is often a prelude to growth. The tension can only be held when we consider our individual and collective trauma from centuries of oppression. Healing must be a priority as we work toward actualizing a society in which every human being can self-actualize and be free from harm.

Discussion

Embracing anti-racist social work praxis would require significant changes in how our discipline operates. Manifesting such changes would be far from simplistic as the gap between who we are as a discipline and who we would like to become is vast; however, a few challenges we must immediately grapple with include the significant gap in preparation of our current workforce, the misalignment between the current roles of many social workers and anti-racist social work praxis, and lack of will.

Social work professionals must be prepared to engage critical theories to embrace an anti-racist social work praxis; however, a large number of current professionals may be unfamiliar with these theories (Constance-Huggins, 2012). Those unfamiliar may include social work professionals we typically rely on to educate our larger workforce, including social work educators and professional leaders. Similarly, embracing anti-racist social work praxis requires our professionals to advance macro-level approaches to eliminate structural racism but many social workers' initial education may not have included adequate development of macro-social work practices (McBeath, 2016). In order to move anti-racist social work praxis forward, continuing education must include universal exposure and embracement of critical race theories and macro-practice skills, this content must be incorporated into degree programs' coursework, and training on how to effectively teach this content must be provided to social work educators and leaders. A variety of different entities would need to coordinate efforts to shift social work educational practices in these manners. Which institutions in our profession will ultimately be responsible for developing, funding, implementing, and holding the profession accountable to such a revamp of our educational foundations?

As social workers begin to embrace anti-racist social work praxis, they may conclude that their current roles do not allow them to practice in a manner that aligns with its foundational principles. While we do not entirely dismiss the idea that social workers can be agents of change within organizations that do not embrace anti-racist social work praxis, we also believe it essential for social workers to refuse to continue to support racist practices and policies. Dismantling white supremacy requires divesting from it, including divesting our labor (Stevenson & Blakey, 2021). As some social workers begin to divest from working in immigrant detention centers, police departments, jails, prisons, and child welfare services, they may need support to find and prepare for new positions. In addition, funding and additional supports must exist to encourage the creation of organizations committed to anti-racist social work praxis and positions in which social workers can practice in such a fashion. Our current racial capitalist systems will not adequately fund work that aims to eliminate structural racism (Incite! Women of Color Against Violence (Ed.), 2017), meaning much of the work necessary to live up to the grand challenge may not be fund-able or carried out as part of our formal employment roles. How do we hold this tension of needing adequate compensation to live and have our work valued while also recognizing that much of the deeper structural change work required of us may not be compensated? In this sense, efforts to eliminate racism may be at odds with the hyper-professionalization of social work itself.

While inadequate preparation and misalignment in current roles are significant challenges to the profession-wide embracement of anti-racist social work praxis, the most daunting barrier is will. As members of a society that has embraced white supremacy so fully, untangling ourselves from its grips is a daunting task that requires constant attention. While current events have turned many members of our profession's attention to racial justice issues, only long-term commitment could lead to the implementation of anti-racist social work praxis as we have proposed in this paper. Additionally, social workers have benefited from white supremacy shaping our profession to fit within its bounds; thus challenging white supremacy will undoubtedly require members of our profession to forfeit power and privileges to which they have become accustomed. Social workers must develop the will to begin and sustain anti-racist praxis even through the challenges, risks, punishments, and danger that often accompany racist backlash. How do we build the political, psychological, and spiritual will to eliminate racism across the multiracial, white-dominant, and geographically diverse profession of social workers in the United States?

Conclusion

The adoption of Eliminating Racism as the 13th Grand Challenge of Social Work presents us all with important opportunities in our efforts to address structural racism. As we contend throughout this paper, our discipline's ability to live up to this grand challenge would be greatly enhanced by replacing the predominant social justice paradigm with principles informed by abolition praxis. Abolition requires us to situate our work in its broader historical context, to constantly engage in the praxis of critical reflection and action, and to center those targeted by white supremacy. Abolition calls on us to reconceptualize our anti-racist efforts with more specificity and to be vigilant about the ways well-intentioned change efforts can reinforce structural racism.

In this paper, we offer a conceptual framework, principles, and guiding questions to inform anti-racist social work praxis, as informed by abolition. All anti-racist social work praxis needs to take as its central aim the building of power in Black, Indigenous, or Brown and poor communities. The additional principles of engaging critical theories, advancing macro-level approaches, targeting racism at the source, and developing anti-racist interventions are also critical components of comprehensively working toward the elimination of racism. Racism is everywhere; to work toward its elimination we must build specific, proactive strategies.

The work ahead of us will not be easy. While Black, Indigenous, or Brown and marginalized communities continue to lead resistance to white supremacy and efforts to eliminate structural racism, none of us entirely know the answers. In addition to the framework and guiding questions offered here, we provide tools for groups to engage in anti-racist praxis through a framework of B.I.G., recognizing our [B]ias, setting Intentions, and co-creating Ground rules. This tool can be utilized to support groups in working through the various challenges and resistances we will encounter as part of our active participation in the generations-long struggle to eliminate white supremacy. We offer this work in honor of the racial justice, liberation movements before us and in service to the anti-racist work we hope to see social work authentically, accountably practice in the years to come.

Note

1. After pushback from faculty and students around the country, in August 2020 CSWE launched a taskforce to Advance Anti-Racism in Social Work Education. Drafts of the upcoming EPAS may address the unacceptable absence of anti-racist practice in the standards.

Acknowledgments

The authors would like to thank the racial justice and abolitionist movements that have provided many of the insights reflected on in this paper. We have no known conflicts of interest. Data sharing is not applicable to this article as no new data were created or analyzed in this study.

Disclosure statement

No potential conflict of interest was reported by the author(s).

ORCID

Kristen Brock-Petroshius ⓘ http://orcid.org/0000-0003-4714-5277

References

Abdulle, A., & Obeyesekere, A. N. (Eds.). (2017). *New framings on anti-racism and resistance: volume 1 – Anti-racism and transgressive pedagogies*. Sense Publishers. https://doi.org/10.1007/978-94-6300-950-8

American Radio Works. (1966, October 29). *Stokely Carmichael | Speech at University of California, Berkeley*. Say It Plain, Say It Loud: A Century of Great African American Speeches. http://americanradioworks.publicradio.org/features/blackspeech/scarmichael.html

The Associated Press. (2020, June 7). Video: Minneapolis City council pledges to disband police dept. *The New York Times*. https://www.nytimes.com/video/us/politics/100000007179116/minneapolis-city-council-police.html

Barth, R., Gehlert, S., Joe, S., Lewis, C. E., Jr., McClain, A., Shanks, T., Sherraden, M., Uehara, E., & Walters, K. (2019). *Grand challenges for social work: Vision, mission, domain, guiding principles, & guideposts to action* (pp. 4). American Academy of Social Work and Social Welfare.

Bell, J. M. (2014). *The black power movement and American social work*. Columbia University Press.

Bonilla-Silva, E. (2010). *Racism without racists: Color-blind racism and the persistence of racial inequality in the United States* (3rd ed.). Rowman & Littlefield Publishers.

Brown, A. M. (2020). *We will not cancel us: Breaking the cycle of harm*. AK Press.

Carruthers, C. A. (2018). *Unapologetic: A black, queer, and feminist mandate for radical movements*. Beacon Press.

Constance-Huggins, M. (2012). Critical race theory in social work education. *Critical Social Work*, *13*(2), 1–16. https://doi.org/10.22329/csw.v13i2.5861

Cosgrove, J. (2020, November 3). *L.A. County voters approve measure J, providing new funding for social services. Los Angeles Times*. https://www.latimes.com/california/story/2020-11-03/2020-la-election-tracking-measure-j

Council on Social Work Education [CSWE]. (2015). *Educational policy and accreditation standards for baccalaureate and master's social work programs* (pp. 20). https://www.cswe.org/getattachment/Accreditation/Accreditation-Process/2015-EPAS/2015EPAS_Web_FINAL.pdf.aspx

Craig de Silva, E., & Clark, E. (2007). Institutional racism & the social work profession: A call to action. In *President's initiative: Weaving the fabrics of diversity* (pp. 32). National Association of Social Workers [NASW]. https://www.socialworkers.org/LinkClick.aspx?fileticket=SWK1aR53FAk%3D&portalid=0

Critical Resistance. (2020). *Reformist reforms vs. Abolitionist steps in policing.* https://static1.squarespace.com/static/59ead8f9692ebee25b72f17f/t/5b65cd58758d46d34254f22c/1533398363539/CR_NoCops_reform_vs_abolition_CRside.pdf

Davis, A. Y. (2005). *Abolition democracy: Beyond empire, prisons, and torture* (Seven Stories Press 1st ed.). Seven Stories Press.

Dettlaff, A., & Abrams, L. (2020, June 23). An open letter to NASW and allied organizations on social work's relationship with law enforcement. *Medium.* https://medium.com/@alandettlaff/an-open-letter-to-nasw-and-allied-organizations-on-social-works-relationship-with-law-enforcement-1a1926c71b28

Dolan, M. (2020, June 12). *London Breed pushes San Francisco reforms: Police no longer will respond to noncriminal calls. Los Angeles Times.* https://www.latimes.com/california/story/2020-06-12/san-francisco-police-reforms-stop-response-noncriminal-calls

Du Bois, W. E. B. (1935). *Black reconstruction in America: 1860 - 1880* (1. ed.). The Free Press.

Freire, P. (1970). *Pedagogy of the oppressed* (30th anniversary ed.). Continuum.

Gee, G., & Ford, C. (2011). Structural racism and health inequities: Old issues, new directions. *Du Bois Review: Social Science Research on Race, 8*(1), 115–132. https://doi.org/10.1017/S1742058X11000130

Gramsci, A. (1971). *Selections from the prison notebooks* (Q. Hoare & G. N. Smith, Eds.). Lawrence & Wishart.

Grand Challenges for Social Work. (2020, June 26). *Grand Challenges for Social Work Announces Grand Challenge to Eliminate Racism.* /grand-challenges-for-social-work/announcing-the-grand-challenge-to-eliminate-racism/

Hernandez, K. L. (2017). *City of inmates: Conquest, rebellion, and the rise of human caging in Los Angeles, 1771-1965.* The University of North Carolina Press.

hooks, B. (1994). *Teaching to transgress: Education as the practice of freedom.* Routledge.

Hooks, M. (2020, June 23). Policing has failed, we need a million community-driven Experiments. *Essence.* https://www.essence.com/feature/defund-police-abolition-m4bl-song/

Incite! Women of Color Against Violence (Ed.). (2017). *The revolution will not be funded: Beyond the non-profit industrial complex.* Duke University Press.

Jacobs, L., Kim, M., Whitfield, D., Gartner, R., Panichelli, M., Kattari, S., Downey, M. M., Stuart Mcqueen, S., & Mountz, S. (2021). Defund the police: Moving towards an anti-carceral social work. *Journal of Progressive Human Services, 32*(1), 37–62. https://doi.org/10.1080/10428232.2020.1852865

Kaba, M. (2021). *We do this 'til we free us: Abolitionist organizing and transforming justice* (T. Nopper, Ed.). Haymarket Books.

Kelley, R. D. G. (2002). *Freedom dreams: The black radical imagination.* Beacon Press.

Kendi, I. X. (2019). *How to be an antiracist* (First ed.). One World.

Kim, M. E. (2019). Anti-carceral feminism: The contradictions of progress and the possibilities of counter-hegemonic struggle. *Affilia, 35*(3), 309–326. https://doi.org/10.1177/0886109919878276

Kivel, P. (2017). Uprooting racism: How white people can work for racial justice.

Lipsitz, G. (2006). *The possessive investment in whiteness: How white people profit from identity politics* (Rev. and expanded ed.). Temple University Press.

Magill, R. S., & Clark, T. N. (1975). Community power and decision making: Recent research and its policy implications. *Social Service Review, 49*(1), 33–45. https://doi.org/10.1086/643208

Major, D. (2020, September 8). *Democrats introduce bill declaring racism a public health crisis. Black Enterprise.* https://www.blackenterprise.com/democrats-introduce-bill-declaring-racism-a-public-health-crisis/

McBeath, B. (2016). Re-envisioning macro social work practice. *Families in Society: The Journal of Contemporary Social Services, 97*(1), 5–14. https://doi.org/10.1606/1044-3894.2016.97.9

Miller, J., & Garran, A. M. (2017). *Racism in the United States: Implications for the helping professions.* Springer Publishing Company.

National Association of Social Workers [NASW]. (2017). *Social work code of ethics* (pp. 24). https://socialwork.utexas.edu/dl/files/academic-programs/other/nasw-code-of-ethics.pdf

National Association of Social Workers [NASW]. (2021). *2021 Blueprint of federal social policy priorities* (pp. 47). https://www.socialworkers.org/LinkClick.aspx?fileticket=KPdZqqY60t4%3d&portalid=0

Omi, M., & Winant, H. (2015). *Racial formation in the United States* (Third ed.). Routledge/Taylor & Francis Group.

powell, J. A. (2008). Structural racism: Building upon the insights of John Calmore. *North Carolina Law Review, 86*(3), 791. http://scholarship.law.unc.edu/nclr/vol86/iss3/8

Pulido, L. (2017). Geographies of race and ethnicity II: Environmental racism, racial capitalism and state-sanctioned violence. *Progress in Human Geography, 41*(4), 524–533. https://doi.org/10.1177/0309132516646495

Rasmussen, C., & James, K. "Jae." (2020, July 17). Jae. Truthout. https://truthout.org/articles/trading-cops-for-social-workers-isnt-the-solution-to-police-violence/

Reisch, M. (2014). Social justice and liberalism. In M. Reisch (Ed.), *Routledge international handbook of social justice* (1st Edition, pp. 132–146). Routledge.

Reyes, M. V. (2020). The disproportional impact of COVID-19 on African Americans. *Health and Human Rights, 22*(2), 299. https://www.ncbi.nlm.nih.gov/pmc/articles/PMC7762908/

Richie, B. E., & Martensen, K. M. (2020). Resisting carcerality, embracing abolition: Implications for feminist social work practice. *Affilia, 35*(1), 12–16. https://doi.org/10.1177/0886109919897576

Robinson, C. J. (2000). *Black marxism: The making of the Black radical tradition.* University of North Carolina Press.

Roy, A. (2006). Praxis in the Time of Empire. *Planning Theory, 5*(1), 7–29. https://doi.org/10.1177/1473095206061019

Rubinstein, D., & Mays, J. C. (2020, August 10). Nearly $1 billion is shifted from police in budget that pleases no one. *The New York Times.* https://www.nytimes.com/2020/06/30/nyregion/nypd-budget.html

Sinha, M. (2017). The slave's cause: A history of abolition.

Smith, S., Wu, J., & Murphy, J. (2020, June 9). *Map: George Floyd protests around the world. NBC News.* https://www.nbcnews.com/news/world/map-george-floyd-protests-countries-worldwide-n1228391

Spade, D. (2020). *Mutual aid: Building solidarity during this crisis (and the next).* https://rbdigital.rbdigital.com

Stevenson, R. R., & Blakey, J. M. (2021). Social work in the shadow of death: Divesting from anti-blackness and social control. *Advances in Social Work, 21*(2/3), 989–1005. https://doi.org/10.18060/24103

Tai, D. B. G., Shah, A., Doubeni, C. A., Sia, I. G., & Wieland, M. L. (2021). The disproportionate impact of COVID-19 on racial and ethnic minorities in the United States. *Clinical Infectious Diseases, 72*(4), 703–706. https://doi.org/10.1093/cid/ciaa815

Thibeault, D., & Spencer, M. S. (2019). The Indian adoption project and the profession of social work. *Social Service Review, 93*(4), 804–832. https://doi.org/10.1086/706771

Wilkerson, I. (2020). *Caste: The origins of our discontents* (First ed.). Random House.

Power knowledge in social work: educating social workers to practice racial justice

Christopher A. Strickland and Caroline N. Sharkey

ABSTRACT

This article analyzes and interrogates knowledge-production practices in contemporary social work research and practice through the lens of Michel Foucault's concept of power-knowledge. As a regime of power, social work produces forms of knowledge that stratify human subjects along the social fabric. As a result, social work practice and research alike can perpetuate binaries of human existence expressive of the Western context which fashioned it. To reconcile a contemporary social work professional logic saturated in white supremacy with a longstanding ethical mandate for social justice, this investigation concludes with practice and pedagogical recommendations informed by an anti-racist theoretical framework.

In the year 2020, the persistence and pervasiveness of racism were forced onto the U.S. collective consciousness in ways not seen since the 1960s. It was a year defined by a racialized dual pandemic in which historically marginalized U.S. citizens – Black Americans in particular – endured widespread COVID-19 mortality and police-sanctioned violence, both a function of American racism. Black Americans, Indigenous Americans, and Pacific Islander Americans were approximately 3.5 times more likely to die from COVID-19 in 2020 compared to their White counterparts (Gross et al., 2020). Correspondingly, U.S. citizens slain by U.S. police in 2020 surpassed the previous years' rates, with Black Americans experiencing the highest rate of police-involved fatalities in comparison to all other demographics (Statista Research Department, 2022). One such fatality, the murder of George Floyd, Jr., displayed the mercilessness of racism in raw form, as exhibited in a 9-minute clip in which former Officers Derek Chauvin, James Keung, and Thomas Lane torture and murder Floyd, Jr. in broad daylight.

On June 2, 2021, barely a year after the murder of George Floyd, Jr., the Georgia State Board of Education passed a resolution 11–2 to limit classroom discussions of race, stating "the United States of America is not a racist country, and that the state of Georgia is not a racist state." This resolution is one example of a sweeping reaction nationally following the Trump administration's September 2020 memo banning:

> ...training on 'critical race theory,' 'white privilege,' or any other training or propaganda effort that teaches or suggests either (1) that the United States is an inherently racist or evil country or (2) that any race or ethnicity is inherently racist or evil. In addition, all agencies should begin to identify all available avenues within the law to cancel any such contracts and/or to divert Federal dollars away from these unAmerican propaganda training sessions. (M-20-34, p. 1)

To date, 20 states have proposed legislation that restricts the ways educators can discuss racism and sexism in classrooms, including outright restrictions on discussing critical race theory (CRT), with 16 of these bills drawing language directly from the 2020 federal memo (Schwartz, 2021).

Again, as has been characteristic of the total racial history of the U.S., predominating state apparatuses obscure the impact of structural racism with appeals to universalism and nationalism crystallized in policy (K. W. Crenshaw, 2019; Welsh et al., 2021). Unfortunately, social work literature has implicated the social work profession as a fellow avatar for the invisibilizing of American racism and incubator of white supremacy. Critical theorists and social science researchers have illuminated both the existence and impact of social work's prevailing standard of White normativity in its educational and practice settings (Becker et al., 2021; Constance-Huggins & Davis, 2017; Evans-Winters & Twyman Hoff, 2011). These studies suggest social work, practitioners, educators, and researchers alike can engage in practices wherein certain behaviors specific to the predominant culture are deemed normal, and those that do not fit the white normative mold are deemed pathological (Bussey et al., 2021). Accordingly, social work has had great difficulty working with historically marginalized communities among which traditional modes of social work assessment, scaling, and intervention have not been adequately validated or were found to be pragmatically counterproductive generally (Budhwani et al., 2015; Cook et al., 2017, 2019; Lester et al., 2010) and providing services for historically marginalized communities embattled with the cascading effects of COVID-19 specifically.

Furthermore, narrative transcription spanning the past five years uncovers a social work educational environment that is ill-equipped to prepare our students to be cross-cultural, structural practitioners. In response to the ramifications of longstanding cultural insufficiencies, self-identified Black and African American social work students and practitioners disproportionately carry the weight of the profession's longstanding cultural insufficiencies. Recent research suggests that Black and African American practitioners are more likely to serve high-need populations than their White counterparts, despite more than half of all new social work graduates identifying as White (Salsberg et al., 2017, 2020).

This paper will focus on establishing several things. One, it will embark upon showing how power and knowledge operate in the words of Michel Foucault. Second, it will demonstrate how these dynamics operate in the everyday practices of contemporary social work. To help conceptualize this, we have included a diagram of the cyclical nature of power and knowledge (see Figure 1). According to Foucault (1980), power produces forms of knowledge. The knowledge generated modifies, redistributes, and stabilizes categories of power, thus sustaining itself. Power, however, has its own inherent complex dynamics. It establishes institutions, which act as tools of controlling subjects. This relational power – the ability to produce circuits of "normative being" – run to the heart of social work practice and academia.

Power knowledge

The first precept of Foucault's power-knowledge (1980) that we unpack is the powerful link between the two terms it marries: power and knowledge. In other words, power and knowledge are not seen as independent entities but are inextricably related. Knowledge is always an exercise of power and power is always a function of knowledge. And we can see this in the ways through which social work therapy operates at its essence. Let us take grief work, for example. The guiding premise of therapeutic intervention in bereavement is that there are normal and abnormal responses to death and loss (Chambon et al., 1999). This division operates pragmatically. The binary reinforces the target for therapy: that, deemed by social work professionals, there are pathological responses to painful events. Social work practitioners, by nature of their work, interpret the subjective experiences of grief from normative vantage points and utilize certain methodologies to bring the respective client back into spaces of supposed "normality."

The second conception of power-knowledge includes how the dynamic impacts our professional logic. Power-knowledge not only limits what we do but also opens new ways of acting and thinking about ourselves. Power pervades seemingly mundane practices and ideas which structure our

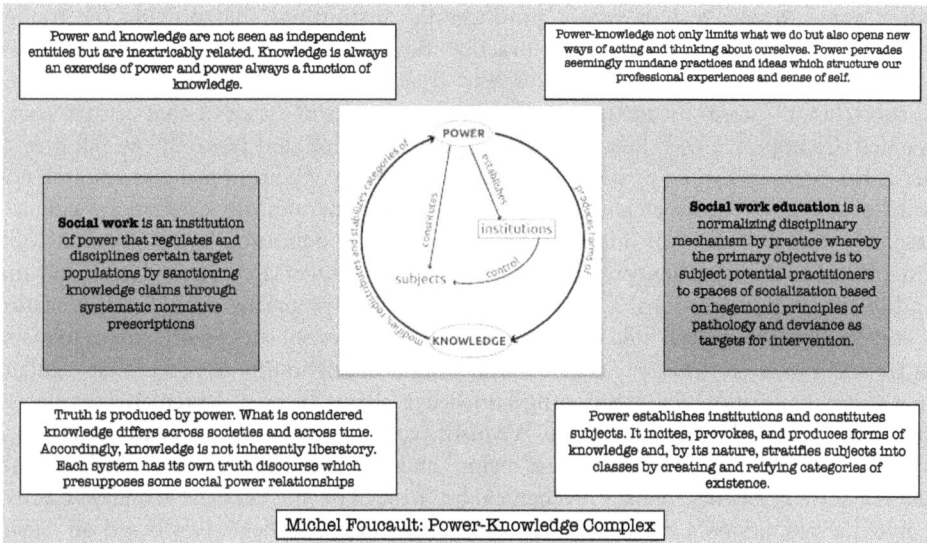

Figure 1. Conceptual model of power-knowledge complex. Adapted from themediastudentblog.com

professional experiences and sense of self. We can see this powerfully in institutions of social welfare, where social workers – their senses of the professional "self" – are anchored in our personas as gatekeepers. The social worker in the contemporary social assistance office has a specific task of acting in the management of the poor and helping to control groups of people who exist at the margin. The "poor ' become the responsibility of a large network of 'social control agents," including the police, parole officers, and social workers. These practitioners develop their own language and knowledge to explain the experiences of poverty, to include distinguishing the "legitimately deserving" from the "deviantly impoverished." While history is riddled with arguments about the rights of individuals as opposed to the moral designation to be more deserving of assistance, "social policies are more widely supported if their beneficiaries are seen as deserving – as having good moral character and being genuinely needy through no fault of their own" (Meanwell & Swando, 2013, p. 495). The clinical relationship relies on a reciprocal relationship guided by the ultimate goal to help, with principles, values, and professional ethical codes to ensure the dignity of the client and the clinician's constant recognition of the client's "worth" (Hebenstreit, 2017). Social workers are encouraged to be aware of the inherent power dynamics that embody the professional role social workers inhabit when interfacing with clients and communities but in practice the role as gatekeepers is defined through the ongoing maintenance of eligibility, criteria, and client "worth" that determine access points to services and resources.

Then we come to the notion of truth, a concept that has always inhabited a constant state of tension and skepticism, but this era of "fake news," these conversations become all the more pivotal. Truth is produced by power. What is considered knowledge differs across space and time. Accordingly, knowledge is not inherently liberatory. Each system has its own truth discourse which presupposes and perpetuates social power relationships. As Scheyett (2006) eloquently states, systematic implementation of evidence-based practices (EBPs) focus little "on the meaning and subjective experience; practices emphasize skills and behaviors, work and education, with little acknowledgement of or attention to the existential challenges and subjective struggles" that are present in the room with the participants of social work programs needing tailored answers to unique issues (p. 75). EBPs are the mascots of the truth produced by social work's institutional power. Embedded in these practices are *a priori* descriptions of the subjectivities of our clients and generalized interventions that further work to objectify the human experience.

We then come to engage how power produces the institutions that provide the framework, privileges, and normative machine for our practice. Power establishes institutions and constitutes subjects (Foucault, 1980). It incites, provokes, and creates forms of knowledge and, by its nature, stratifies subjects into classes by creating and reifying categories of existence that can pervade social work practice (Chambon, 1999). Relatedly, social work education and research can fall prey to this dynamic, as has been uncovered by research done in a plethora of educational and research contexts (Budhwani et al., 2015; Cook et al., 2017, 2019). These dynamics are even more salient when considering the internal issues of homogeneity in social work education and professional settings, norms established by the dominant culture, and disproportionality of clients from historically marginalized and minoritized populations. The stratification of classes permeates social work education and practice and preserve the cyclical hold of racism, sexism, and classism social work strives to dismantle.

What we begin to see conclusively is this: Social work is an institution of power that regulates and disciplines target populations by sanctioning knowledge claims through systematic normative prescriptions. We codify the human experience. We justify colonial modes of intervention in the name of white heteronormative proclamations of "well-being" and "the social good." Accordingly, social work education is a normalizing disciplinary mechanism by virtue of its practice. The primary objective is to subject potential practitioners to spaces of socialization, which are themselves based on hegemonic principles of pathology and deviance. This political designated "Vitruvian Man" becomes our lived ethical code, our successful outcome in program evaluations, the template through which we force groups of the seemingly deviant to fit in order to satisfy our very real role as social control agents in capitalistic society.

Impact of white supremacist logic on social work

Social work's unconfronted white supremacist ideological foundation has created a context where social work educators and higher education administrators have become impediments to anti-racist pedagogy and practice. Faculty are often a barrier to anti-racist pedagogy due to a lack of comfort with the content, which subsequently leads them to avoid including the practice in their teaching methods (Odera et al., 2021; Perez, 2021) or perpetuating naturalized stereotypes of historically marginalized communities in the classroom (Duhaney & El-Lahib, 2021). This includes avoiding conversations regarding the political nature of contemporary culture, Western hegemony, and social work's historic and contemporary role as a steward of white supremacy (Olcon et al., 2020) Correspondingly, White students' level of interest in learning, discussing, and interrogating racism and White privilege has diminished (Abrams & Gibson, 2007; Evans-Winters & Hines, 2020), a demographic that makes up a disproportionate percent of the total social work discipline (Salsberg et al., 2017).

Institutionally, social work administrators who structure social work education enact racist practices and prevent anti-racism from becoming a fixture of the discipline they are tasked to support. Racist hiring practices have often marginalized qualified educators from minority communities (Curiel, 2021). Frequently, Schools of Social Work fail to implement coherent anti-racist policy, curricula, and access to relevant training, intensifying the barrier faculty play to anti-racist pedagogy (Basham et al., 2001; Odera et al., 2021). The multifaceted issue is similarly reflected in how anti-racist pedagogy is typically implemented in social workspaces of education, with less of a focus on how white supremacist culture and its institutions perpetuate racial disproportionality and more so on accepted professional logics of social management and competency development (Jeffery, 2005; Tahir, 2020; Wagner, 2005).

Recommendations

The burgeoning anti-racist praxis that has taken up renewed attention from a wide host of regimes of knowledge in Western society provides the profession a unique way out of the hole it has dug itself in. Anti-racism encapsulates a theoretical and conceptual mosaic, encompassing an extensive spectrum of different perspectives and academic inquiries concerned chiefly with the identification and disman-tling of racism (Anthias & Lloyd, 2002), ranging from Critical Race Theory (Bell, 1987; K. Crenshaw, 1989; Delgado, 1989) to Culturally Sustaining Pedagogy (Ladson-Billings, 1995; Paris, 2012). Anti-racist practice in social work, then, involves wreathing the history, the practices, the identity forma-tions, and the implicit logics of social work interventions within this montage in an effort to rethink why social work operates the way it does today, what aspects of the practice are unsustainable and detached from reality, and an affirmative path to reimaging how to transform a profession drenched in bourgeoise liberalism into an interventionist profession that can appropriately and respectfully engage different registers of human existence.

To address centuries of colonial indoctrination that have plagued the profession and research agenda of social work, practitioners, and educators alike, the profession must begin to interrogate how it produces its knowledges, by whom, and at what costs are inflicted on the profession and our target populations, alike. As it is, the profession must engage in a pre-conceptual deconstruction of itself, the likes of which involves demystifying how the concept power-knowledge operates within its regime of power. It involves developing, not just annual conference spaces through which the profession is intermittently (and tepidly) interrogated, but a consistent framework of deconstructionist thought for practitioners, educators, and researchers alike to utilize consistently throughout and informing of its practices.

As has been suggested by philosophers operating at the margins of postmodernism with a unique vantage point of social work and social sciences alike, social work theory and philosophy does not have the capacity to save itself from the inertia of its historical choices. The profession cannot rethink its practices and reimagine a more socially responsive praxis through lenses crafted by the very regime of knowledge that needs appraisal; doing so ignores the historical underpinnings of social science in general and social work in specific required to sufficiently rethink its theoretical – and metaphysical – orientation (Barad, 2007; Horkheimer, 2002; Zuckert, 1991). We are pressing up against critical mass and to see our profession through to the stated mission to "to enhance human well-being and help meet the basic human needs of all people, with particular attention to the needs and empowerment of people who are vulnerable, oppressed, and living in poverty" (National Association of Social Workers [NASW], 2021, preamble). This requires commitments, stated and acted upon judiciously, to reorient social work education and practice to embody racial justice in all aspects of our profession. Developing anti-racist strategies are the steps necessary to this embodiment, because without racial justice we cannot as a society be socially just.

What does all this lofty re-conceptualization business mean on the ground? From an anti-racist lens, it means shaking off the dews of individualistic neoliberalism that have long made social work the consistent capitalistic beneficiary of a multi-trillion-dollar, biomedical grantmaking enterprise that supports and enables colonial enterprises, focused on the façade of individual successful outcomes, in a reality where the ecological bares deeper on the issues the professions clients face. It means defeating the dominant theologies of free market saviorism and individualism the profession has benefitted from since its inception and engage longstanding issues of racism within the political, socioeconomic, and ideological contexts it manifests its animus from (Macey & Moxon, 1996). It will require a consistent centralizing of targeted marginalized peoples in all levels of social work practice and research, an ethics that requires a tireless dialectic between the interventionist and the intervened (Braidotti, 2013; Graham, 2000).To optimize an archaic, oppressive practice, the community of its practitioners must develop a new calculus of deconstruction to reorient the ontological, epistemological, and methodo-logical presumptions that have been granted unassuming currency through the judicious lens of anti-racist theorems and once reduced to its bare logics, re-built from the ground up with all stakeholders substantively involved at the table. And who exactly is at the table? Rodgers and Lopez-Humphreys (2020) share in their powerful commentary on the Grand Challenges for Social Work Initiative (GCSWI) the crucial need for inclusivity in leadership that shapes the directionality of our profession. They call for GCSWI to embed the outsider-within-leadership (OWL) framework to foster inclusivity by using the very forces of professional logic and subject positionality to affect the change process toward racial justice and more equitable practices (Rodgers & Lopez-Humphreys, 2020). By high-lighting the issues of *who is at the table*, Rodgers and Lopez- Humphreys (2020) hone in on one of the most pressing issues of our capacity as a profession to meet our mission:

> Social policies that obfuscate the underlying structures of inequality, marginalization, andracism (for example, equal opportunity and nondiscrimination or workplace violenceprevention policies) advance practices predi-cated on ideologies embedded in Whitesupremacy (Collins, 2000), promote ongoing institutional racism (Rodgers, 2015, August), andsubvert or negate core social work values. (p. 399)

Social work is at a precipice, with rapidly expanding BSW and MSW programs to meet the demands of a profession that has grown exponentially, with the U.S. Bureau of Labor (2021) projecting the employment of social workers to grow by more than 13% between 2019 and 2029, numbers that may be deeply underestimated considering the acuity of socioeconomic, psychological, and psycho-social impacts as a result of the COVID-19 pandemic and a surge of racially motivated discrimination and violence since February 2020. Yet, the U.S. Human Resources and Services Administration posited prior to the COVID-19 pandemic that there was a national shortage of social workers predicted by 2025 and the basic law of supply and demand illustrate that statistically, we are headed into difficult times to enact slow, progressive change. Rather, we need to act decisively and strip away the holdings of predominantly White institutions and praxis that perpetuates the neutrality and disconnect in our profession.

We need to commit fully to racial justice within our schools of social work and seek to immediately diversify our faculty and student bodies to ensure practitioners and educators represent more fully the racial, ethnic, gender, linguistic, as well as other historically under-represented people in academia. Inclusivity is paramount to growth and the ability for social work as a profession to affect positive change and racial justice. With the increased demands for social workers in the public and private sectors, we need to ensure that we prioritize diversifying the workforce so as to ensure that our profession deconstructs and dismantles the optics of White Saviorism and instead embodies the tremendous power that comes from a workforce that reflects its constituents and the general population at large. We need to encourage greater representation politically and shift the tragic reality that, currently, there are only five social workers in the 117th Congress of the United States as compared to the 320 business professionals and 230 attorneys that are disproportionately represented, this not counting the more than 350 members described as employed in public service/politics (Manning, 2021).

To appropriately prepare new cadres of social work practitioners to address current and future dual pandemics, it is imperative that social work educational construct within its myriad pedagogical settings critically-engaged spaces for professionally self-reflexivity, institutional critique, and political reimagina-tion. It is time for social work, reoriented and realigned, to exemplify anti-racism throughout the very fabric of this profession and become the force for effective social change to push the national conversa-tion about racial justice into the national action for racial justice. Without urgent actions, informed by a more representative group of social workers *at the table* then the projected demands will eclipse the current state of progress. As the poet and activist James Baldwin stated before his death in 1987 more than 30 years ago, "how much time do you want for your 'progress'?"

Disclosure statement

No potential conflict of interest was reported by the author(s).

References

Abrams, L. S., & Gibson, P. (2007). Teaching notes: Reframing multicultural education: Teaching white privilege in the social work curriculum. *Journal of Social Work Education*, 43(1), 147–160. https://doi.org/10.5175/JSWE.2007.200500529

Anthias, F., & Lloyd, C. (2002). Introduction: Fighting racisms, defining the territory. In F. Anthias & C. Lloyd (Eds.), *Rethinking anti-racisms: From theory to practice* (pp. 1–207). Routledge.

Barad, K. (2007). *Meeting the universe halfway: Realism and social constructivism with contradiction.* Duke University Press.

Basham, K. K., Donner, S., & Everett, J. E. (2001). A controversial commitment. *Journal of Teaching in Social Work, 21* (1–2), 157–174. https://doi.org/10.1300/J067v21n01_10

Becker, T. D., Leffler, K. A., & McCarthy, L. P. (2021). Individual characteristics associated with color-blind racial attitudes in master of social work students. *Journal of Social Work Education*, 1–14. https://doi.org/10.1080/10437797.2021.1942352

Bell, D. (1987). *And we are not saved: The elusive quest for racial justice.* Basic Books.

Braidotti, R. (2013). *The posthuman.* Polity Press.

Budhwani, H., Hearld, K., & Chavez-Yenter, D. (2015). Depression in racial and ethnic minorities: The impact of nativity and discrimination. *Journal of Racial and Ethnic Health Disparities, 2*(1), 34–42. https://doi.org/10.1007/s40615-014-0045-z

Bureau of Labor Statistics, U.S. Department of Labor. (2021). *Occupational outlook handbook, social workers.* https://www.bls.gov/ooh/community-and-social-service/social-workers.htm

Bussey, S. R., Thompson, M. X., & Poliandro, E. (2021). Leading the charge in addressing racism and bias: Implications for social work training and practice. *Social Work Education*, 1–19. https://doi.org/10.1080/02615479.2021.1903414

Chambon, A. (1999). Reading Foucault for social work. In A. Chambon, A. Irving, & L. Epstein (Eds.), *Reading Foucault for social work* (pp. 1–292). New York: Columbia University Press. 978-0-231-10717-4.

Collins, P. H. (2000). *Black feminist thought: Knowledge, consciousness, and the politics of empowerment.* New York: Routledge, Chapman & Hall.

Constance-Huggins, M., & Davis, A. (2017). Color-blind racial attitudes and their implications for achieving race-related grand challenges. *Urban Social Work, 1*(2), 104–116. https://doi.org/10.1891/2474-8684.1.2.104

Cook, B., Hou, S. L.-T., Progovac, A., Samson, F., Sanchez, M, Samson, Frank, Sanchez, M. (2019). A review of mental health and mental health care disparities research: 2011-2014. *Medical Care Research and Review 76*(6), 683–710. https://doi.org/10.1177/1077558718780592

Cook, B., Zuvekas, S., Chen, J., Progovac, A., & Lincoln, A. (2017). Assessing the individual, neighborhood, and policy predictors of disparities in mental health care. *Medical Care Research and Review, 74*(4), 404–430. https://doi.org/10.1177/1077558716646898

Crenshaw, K. (1989). Demarginalizing the intersection of race and sex: A black feminist critique of antidiscrimination doctrine, feminist theory, and antiracist politics. *University of Chicago Legal Forum, 1989*(1), 139–167. https://chicagounbound.uchicago.edu/cgi/viewcontent.cgi?article=1052&context=uclf

Crenshaw, K. W. (2019). Unmasking colorblindness in the law: Lessons from the formation of critical race theory. In L. Harris, D. HoSang, & G. Lipsiz (Eds.), *Seeing race again* (pp. 52–84). University of California Press. https://doi.org/10.1525/9780520972148-004

Curiel, L. O. (2021). Interracial Team Teaching in Social Work Education: A Pedagogical Approach to Dismantling White Supremacy. *Advances in Social Work, 21*(2/3), 730–749. https://doi.org/10.18060/24176

Delgado, R. (1989). Storytelling for oppositionists and others: A plea for narrative. *Michigan Law Review, 87*(8), 2411–2441. https://doi.org/10.2307/1289308

Duhaney, P., & El-Lahib, Y. (2021). The politics of resistance from within: Dismantling white supremacy in social work classrooms. *Advances in Social Work, 21*(2/3), 421–437. https://doi.org/10.18060/24471

Evans-Winters, V. E., & Hines, D. E. (2020). Unmasking white fragility: How whiteness and white student resistance impacts anti-racist education. *Whiteness and Education, 5*(1), 1–16. https://doi.org/10.1080/23793406.2019.1675182

Evans-Winters, V. E., & Twyman Hoff, P. (2011). The aesthetics of white racism in pre-service teacher education: A critical race theory perspective. *Race Ethnicity and Education, 14*(4), 461–479. ISSN-1361-3324. https://eric.ed.gov/?id=EJ932889

Foucault, M. (1980). *Power/knowledge: Selected interviews and other writings, 1972-1977.* Random House.

Graham, M. (2000). Honouring social work principles—Exploring the connections between anti-racist social work and African-centred worldviews. *Social Work Education, 19*(5), 423–436. https://doi.org/10.1080/026154700435959

Gross, C., Essien, S., Gross, J., Wang, S., & Nunez-Smith, M. (2020). Racial and ethnic disparities in population level covid-19 mortality. *Journal of General Internal Medicine, 35*(10), 3097–3099. https://doi.org/10.1007/s11606-020-06081-w

Hebenstreit, H. (2017). The national association of social workers code of ethics and cultural competence: What does Anne Fadiman's the spirit catches you and you fall down teach us today? *Health & Social Work, 42*(2), 103–107. https://doi.org/10.1093/hsw/hlx007

Horkheimer, M., O'Connell, Matthew J., & Horkheimer, M. (2002). Critical theory: Selected essays. *Continuum.*

Jeffery, D. (2005). 'What good is anti-racist social work if you can't master it'?: Exploring a paradox in anti-racist social work education. *Race Ethnicity and Education, 8*(4), 409–425. https://doi.org/10.1080/13613320500324011

Ladson-Billings, G. (1995). Toward a theory of culturally relevant pedagogy. *American Educational Research Journal, 32* (3), 465–491. https://doi.org/10.3102/00028312032003465

Lester, K., Resick, P., Young-Xu, Y., & Artz, C. (2010). Impact of race on early treatment termination and outcomes in posttraumatic stress disorder treatment. *Journal of Consulting and Clinical Psychology, 78*(4), 480–489. https://doi.org/10.1037/a0019551

Macey, M. & Moxon E. (1996). An Examination of Anti-Racist and Anti-Oppressive Theory and Practice in Social Work Education. *The British Journal of Social Work, 26*(3), 297–314.

Manning, J. E. (2021). *Membership of the 117th Congress: A profile* (R46705). Congressional Research Service. https://crsreports.congress.gov/product/pdf/R/R46705

Meanwell, E., & Swando, J. (2013). Who deserves good school? Cultural categories of worth and school finance reform. *Sociological Perspectives, 56*(4), 495–522. https://doi.org/10.1525/sop.2013.56.4.495

National Association of Social Workers [NASW]. (2021). *Preamble to the Code of Ethics.* Retrieved June 5, 2021, from https://www.socialworkers.org/About/Ethics/Code-of-Ethics/Code-of-Ethics-English

Odera, S., Wagaman, M. A., Staton, A., & Kemmerer, A. (2021). Decentering whiteness in social work curriculum: An autoethnographic reflection on a racial justice practice course. *Advances in Social Work, 21*(2/3), 801–820. https://doi.org/10.18060/24151

Olcon, K., Gilbert, D., & Pulliam, R. (2020). Teaching about racial and ethnic diversity in social work education: A systematic review. *Journal of Social Work Education, 56*(2), 215–237. https://doi.org/10.1080/10437797.2019.1656578

Paris, D. (2012). Culturally sustaining pedagogy: A needed change in stance, terminology, and practice. *Educational Researcher, 41*(3), 93–97. https://doi.org/10.3102/0013189X12441244

Perez, E. N. (2021). Faculty as a barrier to dismantling racism in social work education. *Advances in Social Work, 21*(2/3), 500–521. SocINDEX with Full Text. https://doi.org/10.18060/24178

Rodgers S. T. (2015, August). Racism. In Franklin C. (Ed.-in-Chief). Encyclopedia of social work online. New York: Oxford University Press.

Rodgers, S. T., & Lopez-Humphreys, M. (2020). Social work leadership: Grand challenges for black women. *Social Work, 65*(4), 397–400. https://doi.org/10.1093/sw/swaa041

Salsberg, E., Quigley, L., Mehfoud, N., Acquaviva, K., Wyche, K., & Sliwa, S. (2017). *Profile of the social work workforce.* Council on Social Work Education and National Workforce Initiative Steering committee. https://www.cswe.org/Centers-Initiatives/Initiatives/National-Workforce-Initiative/SW-Workforce-Book-FINAL-11-08-2017.aspx

Salsberg, E., Quigley, L., Richwine, C., Acquaviva, K., Sliwa, S., & Wyche, K. (2020). *The social work profession: Findings from three years of surveys of new social workers.* Council on Social Work Education and National Workforce Initiative Steering committee. https://www.cswe.org/CSWE/media/Workforce-Study/The-Social-Work-Profession-Findings-from-Three-Years-of-Surveys-of-New-Social-Workers-Dec-2020.pdf

Scheyett, A. (2006). Silence and surveillance: Mental illness, evidence-based practice, and a Foucauldian lens. *Journal of Progressive Human Services, 17*(1), 71–92. https://doi.org/10.1300/J059v17n01_05

Schwartz, S. (2021, June 11). *Map: Where critical race theory is under attack.* Education Week. http://www.edweek.org/leadership/map-where-critical-race-theory-is-under-attack/2021/06

Statista Research Department. (2022). *People shot to death by U.S. police, by race 2022.* Retrieved March 4, 2022, from https://www.statista.com/statistics/585152/people-shot-to-death-by-us-police-by-race/

Tahir, F. (2020). How to (un)-learn cultural (in)-competency in social work: A critical discourse analysis of cultural competency trainings in community mental health. *Social Work & Policy Studies: Social Justice, Practice and Theory, 3*(1), 1–24. https://openjournals.library.sydney.edu.au/index.php/SWPS/article/view/14409

Wagner, A. E. (2005). Unsettling the academy: Working through the challenges of anti-racist pedagogy. *Race Ethnicity and Education, 8*(3), 261–275. https://doi.org/10.1080/13613320500174333

Welsh, M., Chanin, J., & Henry, S. (2021). Complex colorblindness in police processes and practices. *Social Problems, 68*(2), 374–392. https://doi.org/10.1093/socpro/spaa008

Zuckert, C. (1991). The Politics of Derridean Deconstruction. *Polity, 23*(3), 335–356.

Ethical mental health practice in diverse cultures and races

Winnie W. Kung and Sarah Johansson

ABSTRACT

Drawing from Beauchamp and Childress's four ethical principles as an overarching framework, integrating them with the NASW's code of ethics, we examine their intersection with cultural diversity and antiracism, and its implications for mental health practice. We argue that self-determination in collective cultures may involve inclusivity beyond individual clients. Beneficence is culturally defined and evidence-based practices proven effective for some clientele have to be considered together with clients. For non-maleficence, practitioners need to reduce biases and microaggressions to avoid harming clients. Finally, justice is attainable when antiracist approaches are in place and those marginalized have equitable access to culturally-sensitive services.

Introduction

According to the National Association of Social Workers (NASW) code of ethics, the practitioners' ethical responsibilities to clients include the awareness of cultural and social diversity as well as promoting social justice for all people (NASW, 2017, 1.05 & 6.04). Such responsibilities involve understanding the function of culture in human behavior and society, having knowledge and sensitivity to clients' culture in order to provide effective services, and recognizing the nature of social diversity and oppression on the basis of race, ethnicity, gender identity and sexual orientation to mention but a few. Now why is cultural sensitivity so intertwined with ethical practice? It is because ethical practice which upholds human dignity and worth (NASW, 2017) must also recognize equal value of individuals as a member of their own groups without demanding homogeneity; and true multiculturalism negates the imposition of some cultures on others (Taylor, 1997). Insensitivity toward cultural differences and power differentials could lead to cultural oppression and enforcement of standards of the dominant group on minority groups (Sue, 2015). In the Grand Challenges for Social Work, initiated by the American Academy of Social Work and Social Welfare (2022), *Eliminate Racism* is also listed as one of the thirteen challenges that the profession is faced with in order to champion social progress. In health and mental health practice, culture and race are intricately tied to ethics because they affect how practitioners define normality, and how individuals form their "illness explanatory model" which affects their symptom interpretation and expectation of help (Kleinman, 1980). Mental health practitioners' values often shape the goals, intervention approaches, and expected treatment outcome (Henriksen & Trusty, 2005). It is postulated that all systems of psychotherapy or healing is ethnocentric (Sue, 2015). Hence the extent to which practitioners are cognizant of the cultural values and racial reality of their clients and their own biases determines if the treatment process is ethically conducted to meet clients' needs. In this paper we shall employ Beauchamp and Childress (2013) widely cited work on ethical practice in healthcare as the overarching framework to organize our discussion of the consideration of ethical mental health practice while critically examining how facets of the framework need to be expanded to take into consideration diverse cultures and racism.

Beauchamp and Childress (2013) presented four moral principles that lie at the core of ethical healthcare, and these principles have been widely adopted in the biomedical, public health and health policy professions (the book is in its 7th edition). Their principles of respect for autonomy, beneficence, non-maleficence, and justice succinctly captured the key elements of ethical considerations; though developed largely for health care practice they are also relevant in mental health practice as is considered applicable to psychiatry (Singh & Singh, 2016). Beauchamp and Childress (2013) claimed that the framework of their principles is based on "the common/universal morality." It refers to widely shared norms that constitutes right and wrong human conducts seen as universally valid rules that are not specific to cultural groups (p. 3). However, we believe that while central moral norms acceptable to all moral individuals exist, culture and the hierarchical positionality of the practitioner and the client could shape how these ethical principles are applied and their implications. The fact that these principles were developed in the context of the biomedical profession rooted in Western culture makes cultural consideration all the more imperative.

Throughout this paper, we take a broad view on cultural diversity and refer to differences beyond race and ethnicity to include, gender, religion, race, sexual orientation, nativity, ability/disability, socioeconomic status, etc. Since we shall be suggesting some culturally sensitive mental health practices, we would like to first discuss what they may entail. There is no consensus around the definition of culturally sensitive practice in the field, but several associated terms were noted: cultural competence, cultural humility and culturally responsive practice (Chu et al., 2016; Mosher et al., 2017). Among them commonalities are observed. They include a need for the professionals to have cultural *insight and knowledge*, the ability to *emotionally attune* to clients while being aware of one's own biases resulting from the differential in positionality between them and the client, and the *skills to adjust* assessment and interventions to meet clients' needs.

In addressing the issue of racism, we subscribe to Kendi's (2019) position that "there is no neutrality in the racism struggle . . . One either allows racial inequities to persevere, as a racist, or confronts racial inequities, as an antiracist. There is no in-between safe space of 'not racist'" which, in fact, is "a mask for racism." (p. 9). Some believe that the term microaggressions came about to make racism more palatable and is the contemporary form of racism (Kendi, 2019). Microaggression refers to the "brief and commonplace daily verbal, behavioral, or environmental indignities, whether intentional or unintentional, that communicate hostile, derogatory, or negative racial slights and insults toward people of color" (Sue et al., 2007). It includes microinvalidations, microinsults and microassaults, and the harm has been likened to the "death by a thousand cuts" (Sue et al., 2007). These notions could be applied beyond racism to include sexism, genderism, heterosexism and other forms of oppression. Microaggressions refer to the interpersonal interaction, whereas *macroaggressions* are systemic in nature and are seen in programs, policies and practices of institutions and society (Sue & Spanierman, 2020). As many as 53–82% of clients have reported to have experienced microaggressions during therapy sessions (Owen et al., 2014 as quoted in Sue & Spanierman, 2020) thus is of serious concern in mental health service. Culturally sensitive practice is intricately tied to clinicians' ability to detect and counteract micro- and macroaggressions (D'Aniello et al., 2016; Sue & Spanierman, 2020) by taking an actively anti-oppressive stance.

Beauchamp and Childress (2013) argue that the ethical principles of respect for autonomy, beneficence, non-maleficence, and justice that guide practice are not exhaustive, but they illustrate a large territory of the common morality. In the following discussion we shall 1) critically examine the meaning of these four ethical principles and their implications for mental health services in light of diverse cultures and realities for marginalized groups based on the literature and our experiences, and 2) identify some necessary adaptations when applying these ethical principles to ensure the provision of the best culturally sensitive and socially just services at the clinical, organizational and policy levels.

Respect for autonomy

What is autonomy?

Beauchamp and Childress (2013) define respect for autonomy as acknowledging individuals' "right to hold views, to make choices, and to take actions based on their values and beliefs" (p. 106). Such respect is not limited to a respectful attitude, but more importantly, respectful of clients' actions taken. This ethical principle is similar to the notion of self-determination under National Association of Social Workers's [NASW] (2017) code of ethics: "Social workers respect and promote the right of clients to self-determination and assist clients in their efforts to identify and clarify their goals." Beauchamp and Childress (2013) further theorize that autonomous actions are comprised of three conditions – intentionality (of one's own volition), understanding (having the capacity to make decisions), and non-control (unconstrained either by external sources or internal states). In the mental health field, control from external sources may be imposed by mental health practitioners/the systems, and control from internal sources can be due to mental illness. Moreover, the way that we evaluate one's intentionality, understanding and ability to exert internal and external control is at times subject to interpretation of the person in power, often times, the mental health professionals.

Autonomy and cultural diversity

Some critics of the principle of respect for autonomy deemed that it is too individualistically conceived and focuses too narrowly on the self as an independent and rational control agent, neglecting the social context and communal life. These ethicists prefer the notion of "relational autonomy," believing that individuals' identities are shaped through "social interactions and complex intersecting social determinants, such as race, class, gender, ethnicity and authority structure" (Christman, 1995 and Friedman, 2003 as cited in Beauchamp & Childress, 2013). We believe that a narrow individualistic understanding of respect for autonomy is especially unsuitable for collective cultures where the unit to exercise autonomy may be a family rather than an individual such as the Asian culture (Hsieh & Bean, 2014). It means that as practitioners we should take this into account when deciding who to involve in treatment, and the type of treatment modality. For example, family intervention for individuals with serious mental illness may be suitable especially for those from family-centered cultures.

Some cultures may ascribe much respect to professionals' expert knowledge, experience and authority, where clients may prefer more direct advice-giving from practitioners instead of a non-directive approach which encourages self-determination. This is especially important when clients are at a loss of possible options, or are desperate for immediate help in a dire situation after a long treatment delay (Sue & Sue, 2016). The push for autonomy may be frustrating to clients when they need more immediate active recommendations based on the professionals' expertise. We are not suggesting that clients' autonomy should be taken away. However, focusing on clients' autonomy while not taking this into account may be to the clients' detriment. Clients may be more able to self-determine after the crisis situation has subsided, and when they have had a chance to understand the roles of client and practitioner in the treatment process. On occasion, some treatment approaches that involve a teaching role from the practitioner such as family psychoeducation or cognitive behavior therapy may be suitable for some clients.

Sometimes if we uphold the principle of autonomy too far, we may end up disrespecting clients by forcing them to hear information and make choices when they do not want to. While most clients want to be informed, some do not want to know or participate in decision making in a significant way, and some would like to defer to their family members or the professionals. Practitioners need to acknowledge that while we have the duty of respect for clients' *right* to choose, they do not have a *duty* to choose if they prefer to delegate such decisions to others, and we have to respect their autonomy to delegate. In terms of divulging information to clients so that they could exercise self-determination, we may need to take caution when working with native American clients of Navajo decent. In a study it was found that in this culture thought and language are seen to have the power to shape reality and control events, and discussing negative information about the implications of an illness may actually produce such undesirable results (Carrese & Rhodes, 2000). Thus, having had to process such information could be seen as dangerous, and such discussion with clients a violation of their traditions. It was suggested that we should inquire clients' preference in advance on receiving information and making decisions, especially on potentially sensitive topics such as prognosis of serious or life-threatening illness. If clients wish, family members and/or traditional healers could be present in the discussion. Nonetheless, while clients may belong to a certain cultural group cultural homogeneity should not be assumed, and the choice needs to be individual.

Exercise of autonomy when decision-making capacity is impaired

When individuals with serious mental illness are incapacitated so that they do not have a sound mind to understand the implications and consequences of their actions, nor the capacity for internal control, their ability to exercise autonomy may be impeded. According to NASW's ethical principle (NASW, 2017) "social workers may limit clients' right to self-determination when, in the social workers' professional judgment, clients' actions or potential actions pose a serious, foreseeable, and imminent risk to themselves or others." On issues of involuntary hospitalization or psychiatric commitment of seriously mentally ill patients, ethical dilemmas often emerge.

A common example being the imposition of confinement when a client is posing a risk to themselves or others while being in a psychotic state. The practitioner has to balance the client's right to self-determination with the risk to the client and others of any harm that may come as a result of their mental state, which relates to the practitioner's "duty to protect" under the principle of beneficence (Beauchamp & Childress, 2013). Regarding such duties, society as a whole should decide its obligation to act on behalf of the individual and the public in relatively clear terms through policy setting.

An example of the ethical dilemma of valuing client autonomy over the duty to protect is revealed in a PBS's documentary *Frontline* in the episode "Right to Fail" (Jennings & Sapien, 2019) which described the discharge of individuals with serious mental illness from adult homes to community supportive housing after the New York State passed the bill for patients to decide where they live. This policy was beneficial to high functioning individuals who could regain their freedom and even the sense of personhood after years of institutionalization, and clearly reflected a respect for client autonomy. However, without adequate procedures to determine individuals' capacity for self-care and medication administration, and the lack of adequate support such transition proved to be disastrous for many. They fell through the cracks, relapsed, died, abused, or returned to adult residential care eventually. Despite such negative experiences, many individuals with mental illness still prefer to claim their right to live freely in the community and even the "right to fail." We believe that this does not have to be a choice between freedom and humane care. The respect for individuals' autonomy should align with society's duty to protect through adequate care during the transition or for longer term. Careful monitoring by the state of the for-profit care agencies could also prevent many such tragedies.

Client autonomy and racism

Protecting client autonomy should not have unlimited bounds. Ethical decisions around autonomy must consider, not just clients' well-being, but its effect on the larger society (Stirrat & Gill, 2005). This is in line with NASW's code of ethics which acknowledges social workers' ethical responsibilities toward individual clients as well as promoting social justice for all people, "particularly with and on behalf of vulnerable and oppressed individuals and groups" (NASW, 2017). When clients, especially those of the dominant group, disclose racist or other prejudices in therapy, it could cause an ethical dilemma to the practitioners of whether to accept their racist beliefs and protect their autonomy when they have no intention of addressing the issue, or risk confronting it which may be seen as reprimanding clients' personal beliefs thus causing therapeutic rupture. How then does the practitioner balance their antiracist responsibility to the larger community and to individual clients? First, we need to consider the harmful effect of their autonomous decisions of racism and prejudice on the marginalized groups in the community. By keeping silent, the practitioners may be seen as endorsing and validating such ideologies (Drustrup, 2019) thereby participating, reinforcing and perpetuating them. The practitioner has to "address the racism in the room before meaningful change can happen" as stated in the Grand Challenges for Social Work under eliminating racism (Teasley et al., 2021, p. 12). Second, the white racist narrative founded on an idealized version of history and a false sense of conscious or unconscious superiority results in the view of other races being inferior and denigrated, which are all distorted realities that pose a threat to honest understanding of self and others (Drustrup, 2019). When practitioners aim to promote truthfulness and authenticity in the therapeutic process, such distortions need to be confronted and dispelled as we help clients to deal with other defense mechanisms, thus it will be unethical to leave those incorrect and harmful ideologies unchallenged. When we can see racism as a mental health concern, practitioners have the responsibility to help clients to examine its impact on themselves and their relationship with others (Stirrat & Gill, 2005). Just as we employ psychoeducation to work with other clinical issues we can use it in dealing with racism (Drustrup, 2019). Practitioners should gauge when the clinically appropriate moment is to confront clients with the potentially sensitive issue of racism to better engage them, which is usually when the therapeutic alliance is strong and clients feel safe enough to reflect on their conscious or unconscious biases leading to better understanding of self and others (Stirrat & Gill, 2005). In any case, when clients are in the midst of processing complex traumatic problems, it would not be suitable to bring up the issue of racism as it is unlikely to be in the forefront of client's concern and could be seen as a distraction or the practitioners' insensitivity (Drustrup, 2019). To conclude, practitioners need not feel that they are infringing on clients' autonomy as they confront them on issues of racism or other prejudices when we consider them in the light of furthering greater good for the clients themselves and that of the larger community. What should be acknowledged is that this carries a different weight depending on the practitioners' racial and ethnic background, as it is not the role of the oppressed to challenge the oppressors.

Beneficence

What is considered beneficent?

Per Beauchamp and Childress (2013), the principle of beneficence connotes "acts of mercy, kindness, friendship, charity, and the like" (p. 202). Health professionals have moral obligations to act for the benefit of individuals they serve as well as responsibilities to larger society (NASW, 2017). However, they could be put in a bind since what is considered universal good, upon examination, could be "culturally affected perceptions of well-being" (Napier et al., 2014, p. 1608). It is contended that traditional counseling and psychotherapy are derived from Eurocentric culture which emphasizes on individualism instead of collectivism, universality instead of cultural relativism, and objective empiricism instead of clients' subjective experiences (Sue & Sue, 2016). These values could shape treatment goals and intervention strategies that are insensitive to minority clients' cultures. Further, treatment success may be evaluated based on practitioners' prescribed change direction instead of clients' subjective evaluation or that of their significant others, when the individualistic culture is subscribed.

For example, in Bowen's family systems theory, higher level of self-differentiation or greater individuation from family is a key goal that practitioners aim to attain in familial relationships (Bowen, 1966). Such notion had also been criticized as gender-biased since the prescribed ideal was normed according to the male for whom the use of the intellect instead of emotions is considered as indicative of a healthier relational and psychological state (Knudson-Martin, 1994). Further, as Olson (2000) suggested, what is considered an "enmeshed" family in his Family Circumplex Model may in fact be within the norms of being appropriately connected among Hispanics, Asians, Amish or

Mormons where such level of cohesion could be functional within these families (Olson, 2000). Such labels could be imposed on minority races and their family dynamics considered dysfunctional. In the African American culture, given the historical experience of enslavement, strong emotional familial tie including extended families are crucial to the continuation of the social institution of family. Without recognizing the historical injustice and the harsh context for survival, practitioners could easily pass inappropriate judgments. Thus, adaptations of psychotherapy theories to fit clients' unique cultural context in treatment is vital.

Another common assumption in psychotherapy is that clients' capacity to openly communicate their thoughts and feelings is something to promote, especially with significant others on difficult but important issues so as to minimize secrecy or avoidance. Open communication is considered beneficial to all parties concerned which would enable acceptance of the situation and maintenance of positive relationships in the long run (Jourard, 1964). In the case of impending death, practitioners tend to encourage families to address the issue to facilitate anticipatory grief which is believed to provide an easier transition for both the dying person and the relatives (Simon, 2008). However, some cultures may prefer to cope differently, and would rather not explicitly address the impending death. The elderly father with terminal cancer, of the first author's Asian friend, asked the doctor to explain his medical condition to the children and implicitly asked not to be told. The family respected his desire and was able to prepare for the parting through more frequent family gatherings and together visiting places that were meaningful to the father, and it worked very well for them. Coming out as gay or lesbian amidst conservative cultures is another example. The first author knows of a case in which the Asian son introduced his lover to the family as a friend, who visited the family frequently and was very well received. When the couple broke up and the partner stopped coming, the mother was able to comfort the despondent son without explicitly acknowledging the gay relationship. Mental health professionals who are not culturally sensitive may consider the "issues" not being resolved in the most beneficial way, yet their coping matches the families' culture and needs.

Beneficence and evidence-based practice

To be beneficent, social workers have the responsibility to provide the most effective mental health treatments that ameliorate clients' emotional suffering. Such consideration requires us to address the use of evidence-based practice (EBP). It is argued that only when we provide proven effective treatments to clients are we keeping our ethical standard to have their best interest at heart (Klerman, 1990). It has been emphasized that in implementing EBP, clients' culture, characteristics, and preferences should be considered, with decisions made in collaboration with clients (Gambrill, 1999). However, wholesale adoption of EBP has been reported in agencies requiring the use of cognitive behavioral therapy to treat depression, and health care plans only reimburse this approach to ensure cost effectiveness. These practices among agencies and health insurance companies are often mutually reinforcing thereby limiting client and practitioner choices.

Moreover, controversies over EBP abound despite its prominence in the field. One contention is that most EBP established through randomized clinical trial (RCT) studies were based on the mainstream white population, which render their questionable external validity to almost half of the U.S. population, and there is a paucity of studies on models specifically developed for and conducted among minorities (Sue, 2015). In evaluating RCT studies, concern was raised about the focus we place on the internal factors of "active ingredients" in the models while clients' external circumstances, stressors and history are rarely parsed out (Davidson & Hauser, 2015). As social workers, we subscribe to the person-in-environment understanding of clients' conditions, and we often have to deal with "complex social problems [that] do not lend themselves to narrow and discrete interventions" (Gitterman & Knight, 2013, p. 71). The complex social issues are plaguing more prominently clients of minority races and marginalized groups whose limited external resources often hamper their functioning. Interventions to advocate for their needed resources individually and system-wide are just as important, if not more, compared to facilitating psychological changes within the individuals to attain positive treatment outcome. This again sheds doubt on the external validity of some EBP studies.

Furthermore, some literature on potentially harmful therapy (PHT) alerted practitioners of the probable/possible harm in using established treatments on clients that may not fit them (Davidson & Hauser, 2015; Sue, 2015). Proponents of multicultural psychotherapy also contend that empirical findings in outcome studies with objective measures may not be as important as clients' subjective experience in the treatment process and their outcome evaluation (i.e. empiricism vs. experientialism in Sue & Sue, 2016). While we are leery of the use of EBP without client input, we do value verifiable treatment outcomes. However, we believe that a wider spectrum of methodology in the evaluation process should be used instead of hailing the RCT as the only gold standard of inquiry. Furthermore, we believe that mental health practitioners have the responsibility to keep abreast of what treatments have been proven effective and from what clientele the evidence came so that they can consider whether to adopt, adapt, or abandon them in consultation with clients.

Nonmaleficence

Beauchamp and Childress (2013) state that the principle of nonmaleficence obligates practitioners to abstain from causing harm or the risk of harm to clients. From the biomedical perspective, these authors focus mainly on physical harm though also acknowledge mental harm under this principle. They define harm as "killing, causing pain or suffering, incapacitating, offending or depriving others of the goods of life" (p. 150). NASW's code of ethics expects social workers to "ensure the competence of their work and to protect clients from harm" through careful judgment, appropriate education, research, training, consultation, and supervision. In the mental health field, the harm caused could sometimes be subtle but at other times significant when intersecting with diversity issues.

Harm caused by ignorance

Due to ignorance or cultural insensitivity practitioners may consider what is normative in a certain group as pathological (Sue, 2015). Practitioners may invalidate or ignore cultural interpretations of symptoms such as somatization, a common phenomenon taking different forms in different cultures (United States Department of Health and Human Services [US DHHS], 2001). Some manifestations of mental illness like ghost sickness, heartbreak syndrome, and culture-bound syndromes are common among American Indian and Alaskan Native populations (Manson et al., 1985). Some culturally endorsed indigenous or alternative mental health interventions are sometimes dismissed as unscientific. An example is the use of traditional Chinese medicine or acupuncture to reduce anxiety, depression and insomnia which has been found effective in rigorous studies and reviews (e.g., Leo et al., 2007). Such nonrecognition or misrecognition not only harms the therapeutic alliance, it is, in fact, a form of microaggression, invalidating the minority clients' belief systems and signaling their need to assimilate to the dominant culture (Sue et al., 2007). An inferior or demeaning image of another person can distort and oppress when imposed by the more powerful party, especially if the image is internalized (Sue & Sue, 2016). In the mental health setting, the practitioner is clearly in a position of power.

When the dominant culture inhumanely suppresses individuals' identities, and seeks to mold them according to the dominant culture, such as the use of sexual orientation conversion therapy, clear harmful effects have been reported (Schroeder & Shidlo, 2008). Based on dominant cultural values and ignorance, justified by misinterpretation of phenomenon with biased sampling had led to pathologizing homosexuality, which was removed from the Diagnosis and Statistical Manual of Mental Disorder 2nd Edition (DSM-II) only in 1973. Propositions of reparative treatment is still noted nowadays causing much harm (Drescher, 2015).

Harm caused by bias and insensitivity

When practitioners assume the position of color-blindness and impartiality in race-related issues, they are, in fact, oblivious to the lived experience of racial discrimination among clients of racial minority. A common form of microaggression in therapy is the invalidation of clients' experience of being oppressed (Sue et al., 2007). An example is when clients share how they were treated differently during certain interactions due to their race, practitioners, especially those of the dominant group, may reject the idea, suggesting that clients are overly sensitive which causes their avoidant behavior. Hence, the treatment goals are to identify the "root" cause of "oversensitivity" and to acquire coping skills in future interactions. This approach of pathologizing individual clients and ignoring the pathology of the larger discriminatory and oppressive context is at the very least a form of microaggression, if not blatant racism. The practitioner is causing further harm to the client by adding insult to injury. Such insensitivity and bias could appear in discrimination in terms of gender, sexual orientation, immigrant status and a whole host of membership to marginalized groups.

Only when practitioners are able to confront their own biases would they be able to maintain impartiality in discerning the dynamics of clients' experiences and the context in which they live. To do so, self-reflection on one's own attitudes toward their own racial identity is the first step which echoes the first strategy to eliminate racism under the Grand Challenges for Social Work at the individual level (Teasley et al., 2021, p. 11). It was found that individuals' less integrated forms of racial identity are associated with their higher tendencies to minimize or distort the existence of contemporary racism and "denigration and hostility towards people of color," and was also associated with less racial openness (Gushue & Constantine, 2007). For effective therapy to occur, the dominant group of white practitioners have to be aware how their race and culture "have various biases and assumptions that influence their worldviews and therapeutic work" (Drustrup, 2019). It is of paramount importance that practitioners recognize clients as experts of their own oppression in order to establish trust so as to more fully understand clients' main issues (Sue & Spanierman, 2020). In situations when clients share incidences in which they experience oppression, the best intervention is to empathize and validate clients' experience without judgment, to encourage them to tell their full stories, and collaboratively develop treatment goals that could help them determine personally what the optimal ways are to respond to such provocations. Therapy should be a safe space to process how power and oppression impacts clients' lives (Gushue & Constantine, 2007; Sue & Spanierman, 2020). To create the safe space, we as practitioners need to be able to perceive and address therapeutic ruptures (Davidson & Hauser, 2015), to invite such conversations, and to open the door for clients to share with us when we commit microaggressions toward them and the impact it has on them. The second author recalls a young black woman in one of her early sessions for treatment of anxiety, where the topic of the client's racial identity came up while discussing her symptom development. The author acknowledged her whiteness and highlighted the limitations she may have in fully understanding the client's experience, but invited her to explore this topic together at whatever level she was comfortable. The client started crying, expressing relief over this being acknowledged, and in subsequent sessions not only delved into this topic, but also was empowered to share with this author when her interpretation lacked nuance due to her privileged position. In terms of the therapeutic alliance, inviting this feedback allows for repair and for the clinician to avoid causing future ruptures. It should be noted that it is not the client's responsibility to educate the clinician but it is done based on the client's level of comfort and sense of safety.

The way we as practitioners use language in our interactions with clients can either create safety or indicate to clients that treatment with us is not a safe space. A way to practice cultural sensitivity is to allow the client to share their preferred terminology with us. One example would be introducing ourselves to all new clients with our pronouns and asking for theirs, in an effort to offer safety to non-binary and transgender clients and not risk the harm of misgendering anyone (Knutson et al., 2019). Another example can be asking a client with a disability if they prefer the use of person-first language – person with disability, where it is signaled that their disability is only one aspect of their multifaceted identity, e.g., woman with depression, or if they prefer identity-first language – disabled person, where the belonging to the larger disabled community as a part of their identity is emphasized e.g., depressed woman (Dunn & Andrews, 2015).

Harm caused by misdiagnosis

An even greater significant impairment engendered due to racism and cultural bias is the misdiagnosis of mental disorders. A specific example of racism impacting practitioners' ability to effectively conduct assessments is the tendency to misdiagnose African Americans as having schizophrenia and other psychotic disorders instead of affective disorders (Delphin-Rittmon et al., 2015; US DHHS, 2001). This causes clients to be prescribed neuroleptics with severe side effects for a diagnosis they do not have (Metzl, 2010). Another example is that Asian Americans tend to divulge more somatic symptoms instead of emotional distress compared to white Americans when depressed, which may cause them to go undiagnosed (US DHHS, 2001).

Fortunately, in DSM-IV-TR and DSM-5 (American Psychiatric Association, 2000, 2013), attempts were made to increase practitioners' awareness and ability to correctly diagnose clients from diverse populations by expanding on and refining the variance in how symptoms are described and experienced across cultures. For example, the diagnostic criteria for schizophrenia have noted culture-related issues and caution labeling visual or auditory hallucinations within a religious context (e.g., hearing God's voice), which could be a normal part of a religious experience. In considering "disorganized speech" linguistic variations in narrative styles need to be considered if English is not clients' primary language. Similarly, the Manual cautions practitioners to be sensitive when assessing emotional expression such as eye contact and body language, as these vary across cultures and should be carefully considered before listing them as pathological symptoms. The second author, having moved to the United States from northern Europe in her late 20s frequently received comments about her abnormal eye contact, which is considered too intense by US-born individuals she came in contact with. However, clients from other more reserved cultures such as Asians who usually have less direct eye contact patterns (Akechi et al., 2013) may mistakenly be considered as avoidant or anxious. Pathologizing cultural communication styles this way is a form of microaggression (Sue et al., 2007).

Other cultural considerations to avoid harm

Another cultural consideration in working with clients is around the practice of gift-giving and its implications on professional relationships. Hoop et al. (2008) point out that although gift-giving from clients can be considered a boundary violation, refusing gifts from clients from certain cultures may be more detrimental to the therapeutic alliance and treatment than accepting them, since to the client it might be perceived as a sign that certain emotions are not allowed in therapy. The second author, as a new practitioner, made every effort to be conscious about ethical standards; in her first professional position doing in-home therapy with culturally diverse clients in low-income neighborhoods a mismatch in expectation became evident. As she was offered small items such as food or a snack and drink for the road, initially she always turned them down; but eventually she was told off by an Eastern European client who asked if the worker was trying to offend her since the worker, also being European, "should know that denying tokens such as these is an offense." A similar situation occurred later when a Caribbean client under her breath said something about how nothing she offered was good enough. In these instances, Hoop et al.'s (2008) argument rang true – the practitioner was unable to engage these clients in treatment until she had Russian blinis and Caribbean fruit punch in the clients' homes.

Finally, as indicated in the Surgeon General's Report (US DHHS, 2001), even ethnic minorities who have access to mental health services usually do not have access to culturally and linguistically appropriate services. We would venture to say that the lack of culturally and linguistically appropriate services can be considered harmful. A practitioner who does not understand the client due to differences in cultural values and practices, and/or who is unable to effectively communicate with the client due to linguistic barriers, not only runs the risk of causing offense to the client, but also induces pain or suffering. As we discussed earlier, having one's world view or cultural practices being pathologized, criticized or disregarded may increase the experience of stigma and decrease the odds of the client staying in the needed treatment. The Surgeon General's 2001 report on mental health service use detailed higher levels of dropout rates for ethnic minorities, and is in line with more recent reports (Wendt et al., 2015). Hence it is only through culturally sensitive and antiracist practices that we could sustain clients' continued participation in treatment when they have a need.

Justice

Beauchamp and Childress's (2013) ethical principle of justice involves distributing benefits, risks and costs fairly, with individuals in similar positions being treated in a similar manner. The authors derive this principle from Rawls' theory of justice (1971) which requires that a just system in which conditions of "fair equality of opportunity" be in place in order to ameliorate the unfortunate effects of characteristics for which the individual is not responsible, for example, race, gender, or social status. Based on the fair-opportunity rule, individuals with functional disability, including mental illness, would need mental health care to reach a suitable functioning level in order to benefit from a fair opportunity in life. Disparities in mental health service need to be addressed to facilitate access by those presently not utilizing it despite their need. Social workers are bound by the ethical principle to challenge social injustice and to "pursue social change, particularly with and on behalf of vulnerable and oppressed individuals and groups" to ensure their access to needed services and resources (NASW, 2017). In the Grand Challenges for Social Work, the profession is seen as a critical social and political institution built upon the values of social justice in the U.S. (Teasley et al., 2021).

The very fact that there are clear health and mental health disparities in the US indicates that justice is not attained and equality of opportunities is not in place. First, the disproportionate underutilization of mental health services among minority groups is striking. The Surgeon General's report lamented that "[r]acial and ethnic minorities do not use mental health services at rates comparable to those of whites or in proportion to the prevalence of mental illness in either minority populations or the general population" (US DHHS, 2001, p. 164). More recent epidemiological studies indicate the same unjust trends (Choi et al., 2019). Second, even when the effect of race is taken into consideration, many individuals who are socially and economically disadvantaged do not have access to care since health insurance is often job-based, and those in low-paying jobs do not have health care plans from their employment package. Third, immigrant status limits mental health care access. Those who are monolingual or have limited English capacity have a slimmer chance to find service providers who speak their languages. Since mental health treatment relies heavily on verbal communication between clients and practitioners to assess the mental health condition and to conduct psychotherapy, linguistic problem creates even greater difficulties than in physical health care. Hence the increase in bilingual mental health workers to improve access is important. Fourth, the lack of cultural sensitivity and inclusiveness could also be a disincentive for continuation of services leading to premature dropout.

Considering the impact of environmental macroaggressions, practitioners need to evaluate the messaging of the environment in our organizations and whether or not it is inclusive (Sue & Spanierman, 2020). When a client enters our agencies, what are the visual cues? Is the wall décor only reflective of the majority-group, or is it inclusive of other group identities? Art work, signage and symbols on the walls are important in signaling a safe inclusive space. If we are serving primarily clients of color, "Black Lives Matter" signs can be an affirmative message. Conversely, if serving primarily LGBTQ clients, having rainbow flags as part of the décor can signal a safe space as soon as clients walk in (Rolón-Dow & Davison, 2021; Sue & Spanierman, 2020). Other aspects of the agencies' tangible set-up which affect clients' first impressions include: Are the intake forms inclusive? Is the front desk staff culturally sensitive or are microaggressions happening already in the waiting room? Is there no clinical staff of color?

The second author has had numerous clients share how intake forms at some agencies were a red flag due to such insensitivities, for example, the items offer only binary gender options of male or female, and options of race include only white or nonwhite. Clients have also shared that during the initial intake assessment, practitioners ask questions that are non-inclusive, such as asking a woman who has listed on an intake form that she is married, how she met her *husband*, assuming that she is heterosexual which is a heteronormative microaggression that may signal to a new client that this is not a safe space. Another client shared how a practitioner spent two sessions conducting an intake psychosocial assessment and not once asked her how being a woman of color adopted by white parents and being raised in a predominantly white neighborhood in the conservative South has impacted her. Each of these clients dropped out of therapy after a few sessions and it took them up to a year to brave another attempt at seeking treatment. If a practitioner does not consider the different systems of oppression their clients are experiencing, they may inadvertently contribute to clients' premature dropout from needed services.

Similar trends have been noted in the LGBTQ community. It has been documented that there are higher rates of unmet mental health care needs and untreated depression in trans- and bisexual individuals despite their higher rate of service use compared to their heterosexual counterpart (Steele et al., 2017). In addition, LGBTQ persons have variance in satisfaction in mental health treatment, with those identifying with multiple minority groups being the worst (Filice & Meyer, 2018). Practitioners have to be affirmative to provide a culturally sensitive safe space for these clients using intersectional approaches, in order to meet their mental health needs.

Furthermore, practitioners and mental health organizations may become barriers to just resource distribution when they have inaccurate understanding of clients' needs. One such example is that professionals do not prioritize respite care for family caregivers of persons with serious mental illness from some minority groups such as Hispanics or Asians, mistakenly aligning with the belief that given their collective culture, caregiving is seen as a family responsibility, hence overlook the immense stress specific families are experiencing. Another example at the opposite end of the spectrum that the first author had worked with is that the case manager, aligning with the individualistic culture, placed a single adult immigrant Chinese woman with serious mental illness in a residential facility, and arranged for senior housing for her elderly parents with whom she had a deep bond, when considering the family's care and housing needs. By ignoring and breaking up their strong family tie, such arrangement proved to be difficult for both parents and daughter as they felt isolated in their respective homes, which was aggregated by their language barrier. More importantly, it broke up the supportive family system where the aging parents would gladly provide caregiving and companionship to the client, which is mutually gratifying. In some situations, foster care homes within the Chinese community are sought for patients with mental illness but to no avail since the culture is not very receptive to opening their homes to an outsider for long-term stay, especially for adults with special need. This situation has significant implications for just and effective service provision at both the policy and direct practice levels in that cultural sensitivity toward particular communities and individuals is important as we consider the suitable unit for service provision and how to make it available.

Another example of injustice imposed on clients is court-mandated therapy for immigrant families, which warrants a multitude of considerations as power differentials, hierarchical structures and oppression are at play. When immigrant parents participating in mandated therapy experience discrimination and devaluation, they may internalize the poor-parent labeling, and generate self-doubt about their parental competence. The second author worked as both therapist and supervisor for several years with families mandated to family therapy as part of the youths' sentencing when they were adjudicated as juvenile delinquents in family court. Some of the first questions that the parents ask are: "Why are we here? We weren't the one committing a crime so why are we being punished? What did we do wrong?" The level of cultural sensitivity practiced by the practitioner also affects clients' attitudes toward both therapy and mental health services, in positive or negative ways. With a more culturally sensitive and empathic provider, clients are likely to be more positive and open in their attitudes to use the opportunity to reflect on more effective ways to relate to their children; whereas those who are treated by a practitioner with limited cultural sensitivity would increase their resentment and negative attitudes (Ahn et al., 2014), not to mention the poor treatment outcome. For practitioners to be able to engage these marginalized families in treatment cultural sensitivity, respect for clients, humility, and compassion are key.

Summary, intersectionality and concluding remarks

In this article, we borrowed Beauchamp and Childress's (2013) ethical principles of respect for autonomy, beneficence, non-maleficence, and justice as an overarching framework to critically examine their relevance to ethical practices in mental health service settings. Integrating these principles with the NASW code of ethics and some notions from the Grand Challenges for Social Work on eliminating racism we reflected on how these values, if applied without consideration of cultures, oppressive systems and intersectional identities could become irrelevant, insensitive and oppressive to service recipients, not to mention the ineffective treatment outcome. Throughout the discussion on the inadequacies of some Eurocentric assumptions, we drew from the literature and our own experiences and suggested what relevant, sensitive and just services should look like. We discussed the implications on mental health service delivery at the clinical, organizational and policy levels. For practitioners to be able to provide ethical mental health service, they have to be, first of all, self-reflective of their own orientations and hidden biases, and be aware of diverse cultures, racism, oppression and micro- and macroaggressions in their practice. Respect for autonomy needs to consider inclusivity beyond the individual to significant others such as family or community, especially for individuals from collective cultures. Practitioners may need to be more active in the treatment process if clients require more input and advice from them. Respect for client autonomy should not negate the need to address issues of racism in treatment. Practitioners also need to recognize that the definition of beneficence could be culturally defined, and clients' own solution to problems needs to be respected. Evidence-based practices which were proven effective for some clientele would have to be considered in collaboration with the clients. Practitioners also need to comply with the principle of non-maleficence by avoiding cultural ignorance and biases leading to microaggression through invalidation, misdiagnosis, dismissing alternative treatments, causing rupture in the therapeutic alliance, and even imposing harmful treatments on clients. Finally, for justice to be attained equitable access to culturally sensitive mental health services are necessary so that minorities and other marginalized populations can have fair opportunities in life to find fulfillment and satisfaction.

Throughout our discussion we highlighted some ethical issues that may emerge as the principles intersect with certain dimensions of diversity and racism. Moreover, we acknowledge that these different dimensions could intersect with each other and may cause more issues especially in terms of social injustice for clients who have aggregates of different minority statuses such as being a queer woman of color of low socioeconomic status. As practitioners, vigilance is necessary to be sensitive to all of the systems of oppression, as well as believing and validating those experiencing oppression. Unless providers and agencies are aware of possible biases through the use of exclusive normative language, forms and procedures, and microaggression, we may become a part of the oppressing systems. This includes our agency environment in terms of décor, diversity of staff hired, and cultural sensitivity of non-clinical office staff. If we fail to take this into account, we may render clients more vulnerable by alienating them from the support they need, and depriving them of their entitled services and benefits. On the contrary, if we embrace the multiplicity of clients' identity, and are able to build trusting therapeutic alliance with those who have been systematically and historically marginalized, we can help to increase their sense of social acceptance and visibility (Davids & Mitchell, 2018).

As noted, if we ignore the idiosyncrasies of diverse culture and the racist societal context in our mental health service provision, we may easily be in violation of our professional code of ethics. We hope that through our afore discussion, professionals in the field, whether they are direct practitioners, agency administrators, policy makers, or students-in-training could become more vigilant so as to provide more culturally sensitive and antiracist services, and continue to self-reflect on our potential biases in order to grow personally and professionally. We believe that what is most important is not only being mindful of keeping the ethical standards but to really strive to reflect on our attitudes and values in interacting with clients who are of a different background and orientation than ours. NASW (2017) states that the code of ethics are values and principles to which we aspire and by which our works are judged, but "[s]ocial workers' ethical behavior should result from their personal commitment to engage in ethical practice." The Grand Challenges for Social Work stipulated that to eliminate racism, at the individual level social workers should commit to continuous learning to critically self-examine and be racially conscious, at the organizational level to collectively promote equitable access to resources and opportunities, and at the level of the profession to critically self-reflect and address its own racist research, practice and polices, and to commit to change systems to become antiracist, sustainable and just at the institutional and organizational levels (Teasley et al., 2021, p. 18).

Finally, we believe that if we are able to commit ourselves to ethical culturally sensitive and socially just practices in our own heart and continue to grow we would be able to do what is stipulated in NASW's ethical principle of respecting the dignity and worth of the person:

> Social workers treat each person in a caring and respectful fashion, mindful of individual differences and cultural and ethnic diversity. Social workers promote clients' socially responsible self-determination. Social workers seek to enhance clients' capacity and opportunity to change and to address their own needs. Social workers are cognizant of their dual responsibility to clients and to the broader society. They seek to resolve conflicts between clients' interests and the broader society's interests in a socially responsible manner consistent with the values, ethical principles, and ethical standards of the profession.

Disclosure statement

No potential conflict of interest was reported by the author(s).

References

Ahn, Y., Miller, M., Wang, L., & Laszloffy, T. (2014). "I didn't understand their system, and I didn't know what to do": Migrant parents' experiences of mandated therapy for their children. *Contemporary Family Therapy: An International Journal, 36*(1), 25–40. https://doi.org/10.1007/s10591-013-9291-1

Akechi, H., Senju, A., Uibo, H., Kikuchi, Y., Hasegawa, T., & Hietanen, J. K. (2013). Attention to eye contact in the West and East: Autonomic responses and evaluative ratings. *PLoS One, 8*(3), 1–10. https://doi.org/10.1371/journal.pone.0059312

American Academy of Social Work and Social Welfare. (2022). Retrieved March 14, 2022, from https://grandchallengesforsocialwork.org

American Psychiatric Association. (2000). *Diagnostic and statistical manual of mental disorders (4th edition, text revision).*

American Psychiatric Association. (2013). *Diagnostic and statistical manual of mental disorders (5th edition).*

Beauchamp, T. L., & Childress, J. F. (2013). *Principles of biomedical ethics (7th edition).* Oxford University Press.

Bowen, M. (1966). The use of family theory in clinical practice. *Comprehensive Psychiatry, 7*(5), 345–374. https://doi.org/10.1016/S0010-440X(66)80065-2

Carrese, J. A., & Rhodes, L. A. (2000). Western bioethics on the Navajo reservation: Benefit or harm? *Journal of American Medical Association, 274*(10), 826–829. https://doi.org/10.1001/jama.1995.03530100066036

Choi, S. W., Ramos, C., Kim, K., & Azim, S. F. (2019). The association of racial and ethnic social networks with mental health service utilization across minority groups in the USA. *Journal of Racial and Ethnic Health Disparities, 6*(4), 836–850. https://doi.org/10.1007/s40615-019-00583-y

Chu, J., Leino, A., Pflum, S., & Sue, S. (2016). A model for the theoretical basis of cultural competency to guide psychotherapy. *Professional Psychology, Research and Practice, 47*(1), 18–29. https://doi.org/10.1037/pro0000055

D'Aniello, C., Nguyen, H. N., & Piercy, F. P. (2016). Cultural sensitivity as an MFT common factor. *The American Journal of Family Therapy, 44*(5), 234–244. https://doi.org/10.1080/01926187.2016.1223565

Davids, C. M., & Mitchell, A. M. (2018). Intersectionality in couple and family therapy. In J. L. Lebow, A. Chambers & D. C. Breunlin (Eds.), *Encyclopedia of couple and family therapy* (pp. 1–6). Springer International Publishing AG.

Davidson, M. M., & Hauser, C. T. (2015). Multicultural counseling meets potentially harmful therapy: The complexity of bridging two discourses. *The Counseling Psychologist, 43*(3), 370–379. https://doi.org/10.1177/0011000014565714

Delphin-Rittmon, M. E., Flanagan, E. H., Andres-Hyman, R., Ortiz, J., Amer, M. M., & Davidson, L. (2015). Racial-ethnic differences in access, diagnosis, and outcomes in public-sector inpatient mental health treatment. *Psychological Services, 12*(2), 158–166. https://doi.org/10.1037/a0038858

Drescher, J. (2015). Out of DSM: Depathologizing homosexuality. *Behavioral Sciences (Basel), 5*(4), 565–575. https://doi.org/10.3390/bs5040565

Drustrup, D. (2019). White therapists addressing racism in psychotherapy: An ethical and clinical model for practice. *Ethics & Behavior, 30*(3), 181–196. https://doi.org/10.1080/10508422.2019.1588732

Dunn, D. S., & Andrews, E. E. (2015). Person-first and identity-first language: Developing psychologists' cultural competence using disability language. *American Psychologist, 70*(3), 255–264. https://doi.org/10.1037/a0038636

Filice, E., & Meyer, S. B. (2018). Patterns, predictors, and outcomes of mental health service utilization among lesbians, gay men, and bisexuals: A scoping review. *Journal of Gay & Lesbian Mental Health, 22*(2), 162–195. https://doi.org/10.1080/19359705.2017.1418468

Gambrill, E. (1999). Evidence-based practice: An alternative to authority-based practice. *Families in Society: The Journal of Contemporary Social Services, 80*(4), 341–350. https://doi.org/10.1606/1044-3894.1214

Gitterman, A., & Knight, C. (2013). Evidence-guided practice: Integrating the science and art of social work. *Families in Society: The Journal of Contemporary Social Services, 94*(2), 70–78. https://doi.org/10.1606/1044-3894.4282

Gushue, G. V., & Constantine, M. G. (2007). Color-blind racial attitudes and white racial attitudes in psychology trainees. *Professional Psychology, Research and Practice, 38*(3), 321–328. https://doi.org/10.1037/0735-7028.38.3.321

Henriksen, R. C., & Trusty, J. (2005). Ethics and values as major factors related to multicultural aspects of counselor preparation. *Counseling and Values, 49*(3), 180–192. https://doi.org/10.1002/j.2161-007X.2005.tb01021.x

Hoop, J. G., DiPasquale, T., Hernandez, J. M., & Roberts, L. W. (2008). Ethics and culture in mental health care. *Ethics & Behavior, 18*(4), 353–372. https://doi.org/10.1080/10508420701713048

Hsieh, A. L., & Bean, R. A. (2014). Understanding familial/cultural factors in adolescent depression: A culturally-competent treatment for working with Chinese American families. *The American Journal of Family Therapy, 42*(5), 398–412. https://doi.org/10.1080/01926187.2014.884414

Jennings, T., & Sapien, J. (2019). Right to Fail. *Frontline.* PBS. Retrieved February 26, 2019, from.

Jourard, S. M. (1964). *The transparent self: Self-disclosure and well-being* (2nd ed.). Wiley.

Kendi, I. X. (2019). *How to be an antiracist.* Random House Publishing Group.

Kleinman, A. (1980). *Patients and healers in the context of culture: An exploration of the borderland between anthropology, medicine and psychiatry.* University of California Press.

Klerman, G. (1990). The psychiatric patient's right to effective treatment: Implications of Osheroff v. Chestnut Lodge. *American Journal of Psychiatry, 147*(4), 409–418. https://davidhealy.org/wp-content/uploads/2014/09/Osheroff-1.pdf

Knudson-Martin, C. (1994). The female voice: Applications to Bowen's family systems theory. *Journal of Marital and Family Therapy, 20*(1), 35–46. https://doi.org/10.1111/j.1752-0606.1994.tb01009.x

Knutson, D., Koch, J. M., & Goldbach, C. (2019). Recommended terminology, pronouns, and documentation for work with transgender and non-binary populations. *Practice Innovations, 4*(4), 214–224. https://doi.org/10.1037/pri0000098

Leo, R. J., Salvador, J., & Ligot, A. (2007). A systematic review of randomized controlled trials of acupuncture in the treatment of depression. *Journal of Affective Disorders, 97*(1–3), 13–22. https://doi.org/10.1016/j.jad.2006.06.012

Manson, S. M., Shore, J. H., & Bloom, J. D. (1985). The depressive experience in American Indian communities: A challenge for psychiatric theory and diagnosis. In A. Kleinman & B. Good (Eds.), *Culture and depression* (pp. 331–338). University of California Press.

Metzl, J. M. (2010). *The protest psychosis: How schizophrenia became a Black disease.* Beacon Press.

Mosher, D. K., Hook, J. N., Captari, L. E., Davis, D. E., DeBlaere, C., & Owen, J. (2017). Cultural humility: A therapeutic framework for engaging diverse clients. *Practice Innovations, 2*(4), 221–233. https://doi.org/10.1037/pri0000055

Napier, A. D., Ancarno, C., Butler, B., Calabrese, J., Chater, A., Chatterjee, H., Guesnet, F., Horne, R., Jacyna, S., Jadhav, S., Macdonald, A., Neuendorf, U., Parkhurst, A., Reynolds, R., Scambler, G., Shamdasani, S., Smith, S. Z., Stougaard-Nielsen, J., Thomson, L., … Woolf, K. (2014). Culture and health. *The Lancet, 384*(9954), 1607–1639. https://doi.org/10.1016/S0140-6736(14)61603-2

National Association of Social Workers. (2017). *Code of ethics.* Retrieved June 28, 2019, from https://www.socialworkers.org/About/Ethics/Code-of-Ethics/Code-of-Ethics-English

Olson, D. H. (2000). Circumplex model of marital and family systems. *Journal of Family Therapy, 22*(2), 144–167. https://doi.org/10.1111/1467-6427.00144

Rawls, J. (1971). *A theory of justice.* Harvard University Press.

Rolón-Dow, R., & Davison, A. (2021). Theorizing racial microaffirmations: A critical race/LatCrit approach. *Race Ethnicity and Education, 24*(2), 245–261. https://doi.org/10.1080/13613324.2020.1798381

Schroeder, M., & Shidlo, A. (2008). Ethical issues in sexual orientation conversion therapies: An empirical study of consumers. *Journal of Gay and Lesbian Psychotherapy, 5*(3–4), 131–166. https://doi.org/10.1300/J236v05n03_09

Simon, J. L. (2008). Anticipatory grief: Recognition and coping. *Journal of Palliative Medicine, 11*(9), 1280–1281. https://doi.org/10.1089/jpm.2008.9824

Singh, A. R., & Singh, S. A. (2016). Bioethical and other philosophical considerations in positive psychiatry. *Mens Sana Monographs, 14*(1), 46–107. https://doi.org/10.4103/0973-1229.193075

Steele, L. S., Daley, A., Curling, D., Gibson, M. F., Green, D. C., Williams, C. C., & Ross, L. E. (2017). LGBT identity, untreated depression, and unmet need for mental health services by sexual minority women and trans-identified people. *Journal of Women's Health, 26*(2), 116–127. https://doi.org/10.1089/jwh.2015.5677

Stirrat, G. M., & Gill, R. (2005). Autonomy in medical ethics after O'Neill. *Journal of Medical Ethics, 31*(3), 127–130. https://doi.org/10.1136/jme.2004.008292

Sue, D. W. (2015). Therapeutic harm and cultural oppression. *The Counseling Psychologist, 43*(3), 359–369. https://doi.org/10.1177/0011000014565713

Sue, D. W., Capodilupo, C. M., Torino, G. C., Bucceri, J. M., Holder, A. M. B., Nadal, K. L., & Esquilin, M. (2007). Racial microaggressions in everyday life: Implications for clinical practice. *American Psychologist, 62*(4), 271–286. https://doi.org/10.1037/0003-066X.62.4.271

Sue, D. W., & Spanierman, L. (2020). *Microaggressions in everyday life (2nd edition)*. John Wiley & Sons, Inc.

Sue, D. W., & Sue, D. (2016). *Counseling the culturally diverse: Theory and practice* (7th ed.). John Wiley and Sons, Inc.

Taylor, C. (1997). The politics of recognition. In A. Heble, D. P. Pennee, & J. R. Struthers (Eds.), *New contexts of Canadian criticism* (pp. 25–73). Broadview Press.

Teasley, M. L., McCarter, S., Woo, B., Conner, L. R., Spencer, M. S., & Green, T. (2021). *Grand challenges for social work initiative*. https://grandchallengesforsocialwork.org/wp-content/uploads/2021/05/Eliminate-Racism-Concept-Paper

United States Department of Health and Human Services. (2001). *Mental health: culture, race, and ethnicity. supplement to "mental health, a report of the surgeon general: Executive summary."* Substance Abuse and Mental Health Services Administration.

Wendt, D. C., Gone, J. P., & Nagata, D. K. (2015). Potentially harmful therapy and multicultural counseling: Bridging two disciplinary discourses. *The Counseling Psychologist, 43*(3), 334–358. https://doi.org/10.1177/0011000014548280

Part III

Impact of dual pandemics on special groups and populations

Necessary, yet mistreated: the lived experiences of black women essential workers in dual pandemics of racism and COVID-19

Rachel W. Goode ⓘ, Kevan Schultz, David Halpern, Sarah Godoy, Trenette Clark Goings and Mimi Chapman

ABSTRACT

The COVID-19 pandemic has showcased the United States' reliance on essential workers, or those deemed necessary to continue critical societal functions. Black women remain overrepresented in essential positions and are on the frontlines of two pandemics: COVID-19 and racism. Using a phenomenological research design, we conducted semi-structured interviews to examine the experiences of 22 Black women essential workers navigating these dual pandemics. Salient themes of these experiences included: desire to and fear of protest; navigating extreme emotions; mixed levels of understanding from colleagues; and a rise in blatantly racist confrontations in the workplace. Further reflection is needed to understand the complex dynamics these women faced.

The emergence of the coronavirus disease 2019 (COVID-19) pandemic has strained many economic sectors while showcasing the nation's reliance on *essential workers*, or those who are engaged in work deemed necessary to continue critical functions in the United States (U.S.). The U.S. Department of Homeland Security (2020) defines essential workers as individuals who conduct a variety of operations and services or support supply chains that are vital to critical infrastructure viability. Industries supported by essential workers include community and government operations, food and agriculture, medical and healthcare, and transportation and logistics (e.g., cashiers, child care workers, house-keepers, physicians, social workers, and teachers; National Conference of State Legislators, 2021; Tomer & Kane, 2020; U.S. Department of Homeland Security, 2020). In the U.S., about 62% of the workforce – 90 million individuals – hold essential positions (Tomer & Kane, 2020). Essential workers face emergent threats to their health by the SARS-COV-2 virus and health consequences of pandemic-induced isolation and stress (Gould & Wilson, 2020; Snowden & Graaf, 2021; Tomer & Kane, 2020). Further, many essential workers did not have the ability to work remotely like many other workers, meaning that they could not stay home and manage children engaged in online learning or care for homebound family members at high risk for contracting the virus. Many essential workers therefore experienced increased stress as they needed to quickly secure childcare and home health care for their loved ones – an emotional and financial burden that placed many one medical emergency away from financial instability. Indeed, essential workers have not only experienced increased exposure to COVID-19; they have experienced increased stress related to COVID-19 potential exposure, caregiving, and other risks and insecurities.

Although essential workers have played a critical role during COVID-19, they remain underserved, underpaid, and often without adequate protective measures against COVID-19, including personal protective equipment (Kallick, 2020; Snowden & Graaf, 2021; Tomer & Kane, 2020). The majority

(75%) of essential workers on the frontlines (e.g., bus drivers, nursing assistants, and retail sales-persons) occupy positions that on average pay lower wages than all paid workers ($21.95 v. $24.98, respectively; Tomer & Kane, 2020). The pay gap experienced by essential workers underscores the social inequities and economic injustices that continue to burden low-wage workers and the communities of color where many essential workers live and work.

Essential workers: Black Americans navigating dual pandemics

Essential workers often represent marginalized communities known to experience racial and social inequities based in systems of power, privilege, oppression which are perpetuated by labor markets and lead to further disenfranchisement (Bailey et al., 2020; Bailey & Moon, 2020; Kantamneni, 2020; Snowden & Graaf, 2021). During COVID-19, non-Hispanic Black individuals were significantly more likely than non-Hispanic White individuals to work in jobs considered essential (e.g., transportation, healthcare, food preparation, and cleaning services) and in occupations that placed them at high risk of contracting COVID-19 and infecting their households (Obinna, 2020; Rodgers et al., 2020). Indeed, although Black workers occupy 13% of all jobs across the U.S. economy, they occupy approximately 19% of low-wage, essential jobs, and remained overrepresented in low-wage, essential positions (Kinder & Ford, 2020). For example, among home health aide positions which are deemed essential, 54% identify as women of color and more than half of these women are Black (Frye, 2020). Further, the top occupations held by Black women are nursing assistants, cashiers, and registered nurses: occupations also deemed essential during the pandemic (Frye, 2020). This disproportionate representation of Black adults among essential workers has had grave consequences: in the U.S., COVID-19 mortality rates to date have been highest among non-Hispanic, Black adults (Rodgers et al., 2020).

What is more, these stressors facing Black essential workers during COVID-19 have been further compounded by the increased visibility of and mobilization against anti-Black police brutality and systemic racism exemplified in the murders of George Floyd, Breonna Taylor, Ahmaud Arbery, and numerous other Black Americans. (Silverstein, 2021; Thomas et al., 2020). Beginning in May 2020, protests due to the murder of George Floyd emerged across the U.S. and by October 2020 nearly 1,000 incidents of police brutality against peaceful protestors and journalists were recorded (Thomas et al., 2020). Racism and police brutality came to be understood as an additional pandemic that created additional stress especially among Black Americans.

Little, however, is known about how these dual pandemics have impacted Black women who were essential workers. Without doubt, Black women are frequently employed in positions deemed necessary and essential for the functioning of society, including nursing assistants, home health aides, cashiers, and registered nurses (Frye, 2020). Even before the pandemic, Black women essential workers often experienced intersectional marginalization due to their to race, gender, and sexual identity (Obinna, 2020). These women face numerous oppressions daily, grappling with microaggressions, stereotypes, racism, and negative biases that affect how they are treated at work, the supermarket, and all public spaces. Furthermore, among Black women who were essential workers and single parents, the challenges of finding adequate childcare and/or supporting children through virtual schooling only compounded an already difficult reality (Frye, 2020; Gould & Wilson, 2020). Not only were these women at increased risk for COVID-19, they also faced limited access to COVID-19 testing centers and the risk of eviction and homelessness if they were laid off or furloughed during the pandemic (Obinna, 2020).

This phenomenological study explores the perceptions and experiences of Black women essential workers during COVID-19 and the increasing visibility of protests against anti-Black violence in the U.S. By examining participants' narratives of their own experiences we seek to better understand their efforts to balance the intersecting stressors associated with their racial identity and professional obligations. Documenting these realities validates and honors these women's experiences and strengthens our collective understanding of how Black women's identity and work has affected them during COVID-19. These findings are of particular importance to social workers who serve these clients and will improve our ability to care for them during and after COVID-19.

Methods

Phenomenological approach

Our phenomenon of interest was Black women's intersectional experiences of racism and COVID-19. Phenomenology posits that reality is a product of one's consciousness and that phenomena can be understood by exploring immediate, lived experience (Groenewald, 2004; Padgett, 2017). This approach supports purposive sampling as an appropriate type of non-probability sampling (Welman & Kruger, 1999) that enables researchers to identify participants based on the purpose of the research topic and the participants' relationship to the phenomenon of interest (Babbie, 1995; Greig & Taylor, 1999; Groenewald, 2004; Kruger, 1988; Schwandt, 1997). Given that phenomenological studies aim for depth, a projected sample size of 20–25 participants was deemed appropriate (Padgett, 2017). The thematic analysis of phenomenological interview data is particularly useful for exploring experiences, situations, and conditions surrounding one's experiences, and identify salient themes (Padgett, 2017).

Author positionality

Understanding the positionality of the researchers conducting this study is an important component of the research study process. Authors 1 and 4 identify as Black American women faculty members. To avoid speaking for the data, both authors made a conscious effort to bracket their personal experiences during the course of the study. Authors 2, 3, and 6 identify as White American men and women faculty members. Finally, one author identified as a Latina woman, and is currently a doctoral student. All authors worked collectively to ensure the study was guided by their cultural knowledge and expertise. Authors 1–3 were involved in the analysis of data and met regularly to ensure the participants' voices were not overshadowed by personal and/or political perspectives. Further, they also consulted with the interviewer, who also identified as a Black American woman, to ensure the integrity of the coding for representing the views of the participants.

Sample

Participants were recruited using purposive and snowball sampling strategies via university listservs and through partnerships with urban, community-based agencies. Eligible participants (a) were 18 years of age or older; (b) identified as a woman; (c) identified as a person of color; (d) were English-speaking; and (e) were employed in one of the following industries between March 2020 and August 2020: food and agriculture, emergency services, transportation, warehouses, delivery, healthcare, communications and information technologies, critical manufacturing, government and community-based services, financial, energy, water or wastewater management, chemical sector, or education.

Procedures

The Institutional Review Board at the University of North Carolina at Chapel Hill approved this study. Potential participants who expressed an interest in the study were provided a detailed information guide outlining the study procedures as well as risks and benefits of participating. In total, 22 essential workers who identified as Black American women and who were from the southeastern portion of the United States participated in this study.

Using a phenomenological qualitative methodology, semi-structured interviews were conducted to gather details about the perceptions and experiences of Black essential workers facing the dual pandemics of racism and COVID-19. Interviews were conducted using virtual videoconferencing software. All interviews were recorded and lasted between 45 and 90 minutes. Participating individuals were compensated with a $35 Visa gift card.

After providing informed consent and completing a brief sociodemographic survey, each participant was asked to respond to two intentionally broad questions: (1) "Tell me about your experience as an essential worker" and (2) "Not only are people of color dealing with COVID-19 in a different way, we/they are doing this in a context of police violence, protest, and national discussions of race. What has it been like for you as an essential worker?" Follow up probes were asked but kept to a minimum to allow participants to describe the phenomenon as they deemed appropriate. After each interview, field notes were completed.

Data analysis

Interviews were transcribed verbatim and read at least twice in their entirety by two coders. Significant words and phrases pertaining directly to the lived experience of COVID-19 and racism were identified. We then began formulating relevant themes common across the participants' transcripts. Thematic analysis (Boyatzis, 1998; Guest et al., 2012) was used to develop a list of codes that identified the major conceptual categories in the data. Initially, the two coders read through a subset of the transcripts (i.e., 4 of 22 transcripts) and independently identified major themes. Their respective code lists were then consolidated into a single codebook which was approved by the principal investigator and research team. Refinements to the list of codes were made through two rounds of test coding using Atlas.ti 8, a qualitative data analysis software program that facilitates multi-coder projects. Disagreements in applications of the codes were adjudicated by the coders and the first author, with invalid code applications being removed from the Atlas.ti data files before the final consolidation of the two sets of codings into a single set of codes applicable to all of the transcripts. These results were integrated to capture and describe the phenomenon of racism and COVID-19. We obtained methodological rigor by applying verification, and met this standard by conducting literature searches related to the phenomenological method, by keeping field notes, and by interviewing participants until saturation was reached (Anderson & Spencer, 2002)

Results

Table 1 presents the demographic characteristics of 21 of the participants.

Notably, one participant did not complete the brief demographic survey; therefore, they were excluded from the survey analysis. All participants identified as Black American women, of which two women also identified as Latinx. Participants lived in North Carolina, South Carolina, and Florida, with the majority (90%) living in North Carolina. Further, participants were between the ages of 22 and 54 years, and 12 participants had at least one child. While 19 of the 21 participants had an educational attainment that was an associate's degree, vocational training, or higher, more than one half ($n = 13$) expressed that their income did not meet their financial needs. Two thirds of participants ($n = 14$) reported a gross household income less than $99,999. Table 2 presents their essential industries and professions. The majority of participants were in employed in the healthcare ($n = 8$) and social work ($n = 6$) fields.

We identified four interrelated themes which described the experiences of Black women essential workers experiencing both racism and COVID-19 simultaneously: (a) desire to and fear of protest; (b) navigating extreme emotions; (c) mixed levels of understanding from colleagues; and (d) a rise in blatantly racist confrontations in the workplace. These themes were consistent across participants irrespective of their profession or sociodemographic characteristics. Descriptions of each thematic element are provided below and supplemented with verbatim examples drawn from the interview participants. To protect participant identities, themes are presented as composite narratives.

Desire to and fear of protest

Participants expressed feeling emotionally moved by the killings of George Floyd and Breonna Taylor, and a strong desire to get involved in dialogue about racism and protests against police brutality. Some acknowledged the desire to channel the stress of the pandemic into activism, noting

Table 1. Demographic characteristics of essential workers of color (*N* = 21).

Characteristic	Participants Mean (SD) or *n* (%)
Age (years)	37.95 (8.85)
Female	21 (100)
African American/Black	21 (100)
Latinx	2 (9.52)
English Speaker	21 (100)
Children	
Yes	12 (57.14)
No	9 (42.86)
Number of children	1.43(1.16)
Educational Attainment	
High School (grades 9–12)	2 (9.52)
Vocational/Technical School	1 (4.76)
Associate's Degree	1 (4.76)
Bachelor's Degree	8 (38.10)
Master's Degree	8 (38.10)
Professional School	1 (4.76)
Household Gross Income	
$20,000 to $29,999	1 (4.76)
$30,000 to $39,999	4 (19.05)
$50,000 to $59,999	3 (14.29)
$60,000 to $69,999	3 (14.29)
$70,000 to $79,999	2 (9.52)
$80,000 to $99,999	1 (4.76)
$100,000 to $150,000	2 (9.52)
Over $150,000	4 (19.05)
Preferred Not to Answer	1 (4.76)
Income Meets Financial Needs	
Yes	13 (61.90)
No	8 (38.10)

Note: Due to incompleteness, one participant's survey was excluded from the survey analysis.

that this action curbed their sense of helplessness in light of persistent and often unpunished anti-Black violence in the U.S. Though many wanted to participate in these protests, they also felt uncertain about whether they would because of their many other responsibilities and their desire to protect their mental health and family. Participants also expressed the difficulties of having to balance care for younger children, interactions with colleagues, and their own peace of mind. As one participant put it,

"And I got so much I have to deal with at home and to get through this. Um, I stopped paying attention to the news. Um, my son had asked me about the Black Lives protest and why weren't we participating, and I felt guilty and bad at the same time, but I was also like, you really don't need to see this negative stuff."

Another participant acknowledged how her position as an essential worker influenced her responses to police violence in the workplace and on social media:

"I don't have the position where you can, um, you know, you can't wear certain things to work or have certain things on, so you can't really, you know, broadcast that. And then if you do, and if you brought that on your social media platforms, you still have to be, um, mindful because you may face backlash."

Several participants also acknowledged their fear for the Black men in their lives, especially their sons. These participants reported uncertainty and fear about how the George Floyd case would resolve and visualizing their own sons when hearing about George Floyd's murder.

"Um, it's been very insightful, um, a little bit scary. I have, um, I have an older, uh, older son, 24. Um, and so as a mom, as a Black mom, a mom of Black boys, it's been very, uh, very scary, um, because you don't know that, you know, what's going to happen throughout the day, if they're gonna make it home safe."

Table 2. Essential industries and professions of participants (N = 22).

Characteristic	Participants n (%)
Education or Educational Services	
Preschool Teacher	2 (9.09)
Speech-Language Pathologist	1 (4.55)
Medical and Healthcare	1 (4.55)
Behavior Health	8 (36.36)
Clinical Phlebotomist	2 (9.09)
Medicaid Case Manager	1 (4.55)
Medical Doctor	1 (4.55)
Medical Scribe	1 (4.55)
Registered Nurse	2 (9.09)
Office Administration	1 (4.55)
Administrative Associate	3 (13.64)
Office Manager	1 (4.55)
Practice Administrator	1 (4.55)
Service Industry	1 (4.55)
Cashier	2 (9.09)
Housekeeper	1 (4.55)
Social Work	1 (4.55)
Child welfare	6 (27.27)
Older adults case planner	5 (22.73)
Other	1 (4.55)
Quality Coach	1 (4.55)

Participants also acknowledged their mixed feelings watching violence that was unfolding as individuals were looting and damaging Black communities. While recognizing that the lived pain in the community was palpable, participants sometimes struggled with a wide range of responses to protesters' expressed rage and actions, including those who took advantage of the situation to destroy property:

> *"And I had mixed feelings of the looting and the—you know, the protesting and the looting. And again, you're watching news, watching different news stations. They have different things to say the commentators have different views and different feelings about what's going on. And, you know, initially I said, why are we burning down our communities? These are our communities. Like I saw buildings and shops and schools being burned down in our Black communities. And I did not understand why are we damaging and burning down our communities, but the rage is so is there, like, if people are just reacting to the rage, they're reacting to the rage."*

Navigating an emotional outpouring

Participants expressed a range of complicated emotions in response to their growing recognition that racism represented an additional pandemic concurrent with COVID-19. They described the present season of dual pandemics as an awakening to systemic racism; one participant described this reality as "overwhelming." Participants described their grief as they watched COVID-19 disproportionately take the lives of Black Americans. Recognizing the additional toll of police brutality on Black Americans further flooded participants with strong emotions:

> *"I feel like as this—yeah, as an essential worker, as anybody of color right now, essential worker or not, um, but just everything that's happening in the world, as far as racial injustice and to have COVID, so police brutality they're killing us, but now we have this whole pandemic that is also killing us because we have—like, African Americans have so many pre-existing conditions that we're at a higher risk of dying due to COVID, more than any other race. So it's like, I don't know. I just feel like it's a lose/lose, like where's the winning? Yeah. It's just too much death."*

Alternatively, other participants reported feeling hopeful that the increased awareness of police brutality by more White Americans would possibly lead to positive change. Participants used phrases such as "now we're going to be woke" and described more people being able to "walk in our [i.e., Black

Americans'] shoes." Participants reported that this perceived increase in empathy was comforting, as was the ability to have more "real" or honest conversations with non-Black individuals. Several participants acknowledged their belief that the world being "shut down" due to COVID-19 was actually positive contributor to getting the attention of many Americans who did not otherwise pay attention to racism:

> "So, um, in the midst of COVID-19 and everything else that was going on around us, I see it as a blessing because people are forced to sit down now and watch TV. You don't have an escape. You don't have anywhere to hide or run you get to see a taste of our reality. And now we can have real conversation and people really, unfortunately, they can start paying attention to taking the knee or police brutality until enough white folk sat down and paid attention to it."

Mixed levels of understanding and responses from colleagues

Participants reported that navigating work environments, especially among predominantly White colleagues, was particularly challenging. Participants described their difficulties speaking openly about police violence and protests, and sometimes feeling that their employers were not addressing the issue directly and feeling frustrated that these issues were "overlooked" and "brush[ed] under the rug." Several participants also noted the lack of understanding from their employers and coworkers and, at times, experiencing deeply alienating encounters with their colleagues.

One participant described a situation in which their clinic had created a space for all employees to discuss race. While the space was well-intentioned, the outcome left the participant feeling "bothered" and "suffocated" by their colleagues' responses:

> "And there was a group of nurses that . . . I found out had went to management and wanted to know why we had to have this special talk for Black people . . . I found myself suffocated by the fact that I had to continue to work with these people, knowing what they had done and, you know, trying to remain professional and unbiased. But knowing that they had these very strong opinions about why do Black people have to have a conversation? You know? And they were very—it was—they were very opposed to it. Um, and they, they tried to escalate it all the way up to HR. That's how against it they were. And that really bothered me because these were three nurses that I work with every single day."

Participants described their discomfort attempting to educate their White coworkers about Black American culture, their own "blackness," the reasons for the protests, and the Black Lives Matter movement. Despite their efforts to educate their White colleagues, participants did not always feel that their coworkers understood. Indeed, though some coworkers would verbally express empathy and outrage about police violence, participants stated that these coworkers' actions did not always match their words:

> " . . . it is a challenge, um, because I feel like some people don't get it. They don't get the reason of the protests, um, they don't get the concept of the Black Lives Matter. Um, they just don't understand it. I was . . . taken back some, um, our clinical coordinators. She had a sign that said All Lives Matter. And all of the women of color that at the clinic with me were like, what? That's not what we're talking about. You know? And I'm like I am not about to argue with you because we're at work. I said, but we had this conversation so many times, I feel like you just don't get it, you know? And that kind of makes me upset."

By contrast, other participants were comforted by the "rise to the call" that others were exhibiting in their workplaces. Participants either reported working in organizations where most of the employees identified as Black or organizations that took the opportunity to make explicit their stance as an anti-racist organization, creating space for "very comfortable" conversations and honest expressions of emotion. As one participant reported,

> "And I remember probably the week after George Floyd got murdered my boss at the time . . . asked me . . ., how are you doing? And I said, do you really want to know? And she said, yeah, I really want to know. And she's white. And I said . . . I'm mad. I'm hurt. I'm upset. I said I'm disgusted that here again, we have to watch on prime TV a Black man get murdered, and I have to still come up in here and deal with the majority of white people and act like it doesn't bother me. And she was like, you don't, you don't have to act like it doesn't bother you. And for me, that was so therapeutic."

Rise in blatantly racist incidents in the workplace

Participants were often overwhelmed by responding to patients' and/or coworkers' overtly racist statements. Indeed, several participants reported confrontations with patients who negatively responded to Black Lives Matter signage in the workplace and/or who explicitly wore masks or shirts with phrases such as "God, guns, and Trump" or "Trump 2020." Participants acknowledged hearing racial slurs used more frequently and/or experiencing increased incidents of disrespect:

> "It's been interesting . . . if you want to go through the last four years have been one thing, um, and they've been worse for me at work and you, but with people trying to spit in my face, because I am Black, uh, the N word flies pretty freely at me while I'm at work. Um, um, COVID is just funny now, right? And like, I do my job well, I'm impartial. Um, you know, it just, in my head, I'm thinking, you're complaining to me about how you don't get enough government support, but then you're wearing this mask that tells me exactly that you voted for the situation that you're currently in."

These incidents often made it challenging for participants to focus on performing their jobs well, and several expressed feeling that they had to "juggle" or manage the need to exhibit professionalism alongside feelings of anger and rage at the pervasive mistreatment of Black Americans. Participants reported having to "push their feelings to the back burner" and work to treat those at work with impartiality despite their clear biases.

> "You get so caught up in . . . the television, news, social media and the hateful things that are being put out and said and done. It gets to you, you know, because this is happening to your people is how you feel. Um, but sometimes you have to turn it off so that you cannot be contaminated by it."

Discussion

The findings from this phenomenological study seek to amplify the voices of Black women essential workers while navigating the twin pandemics of racism and COVID-19. Moreover, these findings reveal the experience of these women manifests in several salient ways: a desire and fear to protest, navigating an emotional outpouring, mixed levels of understanding and empathy from colleagues, and a rise in blatantly racist incidents in the workplace. It is critical to look at the interconnectedness of these themes; per the phenomenological method, disentangling them would tether the meaning attached to this unique experience.

This study's participants – Black women essential workers from a range of employment backgrounds – described their experiences of seeing systemic racism embodied and exemplified during the summer of 2020 by the murders of George Floyd and Breonna Taylor, which brought this issue to the center of a national dialogue. Their interviews and the quotations presented in this paper highlight the complexity of having to work in person in a workplace where conversations were happening about racism. Although some described a hope for progress if White individuals were able to become more knowledgeable of racism, many described interactions that were contradictory (i.e., verbal comments not being reflected in personal actions) or openly hostile, including racist encounters between themselves and White coworkers. These essential workers also expressed wanting to participate in social justice protests but choosing not to do so because they were already stressed by the realities of their work and the COVID-19 pandemic. Having to choose between taking action or protecting their emotional well-being represented a difficult choice with no clear answer. Interestingly, these participants did not speak directly about their fears of getting COVID-19 if they were to attend a protest; instead, they focused on how much mental and emotional capacity they had available at that time when deciding whether or not to attend.

Black American women have been uniquely impacted during the COVID-19 pandemic. Not only are they disproportionately employed in professions identified as "essential," but they are also managing decades of experiences of structural and interpersonal racism and discrimination which, together, increased their vulnerability and susceptibility to COVID-19 (Chandler et al., 2021; Obinnaa,

2020). Per the results of this study, our participants indeed felt vulnerable while balancing the stresses of motherhood, employer expectations, and personal fears about dual pandemics that were ravaging their communities. In another sample of Black women, 40% reported their job was negatively impacted by COVID-19, and 37% acknowledged knowing someone who tested positive (Gur et al., 2020). Some of the stress and worry our participants were facing may have also exacerbated mental health concerns; recent evidence indicates that nearly 14–16% of Black women may have also been managing a diagnosis of depression or anxiety during the pandemic as well (Gur et al., 2020).

Our findings highlight the emotional toll that these twin pandemics of racism and COVID-19 have had on Black Americans, exacerbating worries about physical safety. Isolation imposed to combat COVID-19 may have also foreclosed interactions between coworkers who could have otherwise been important sources of support for one another. This appeared to be true for some participants, but for others, this potential support source in fact became an additional stressor due to mixed messages or openly racist encounters and attacks from coworkers. Furthermore, the cascade of emotions that our participants encountered was overwhelming, with many reporting feeling uncertain about the appropriate response to the protests and national reckoning with police violence. To be sure, this study encourages further reflection on the complex and layered dynamics these women faced.

Limitations

The phenomenological methodology used in this study provides an opportunity to better understand Black women essential workers' experiences of the dual pandemic of racism and COVID-19. Though ours was a purposive sample, the small sample size may limit the generalizability of this work beyond the participants in this study. Moreover, participants were only from the southeast U.S., and their experiences may not represent the experiences of Black women essential workers in other regions of the country. Despite these limitations, this is the first study to (a) explore the experiences of Black women essential workers during the dual pandemics of COVID-19 and racism and (b) highlight experiences that are likely common to many women of color who are essential workers.

Implications for social work practice and policy

These findings have implications for mental health practitioners and administrators, for those trying to create equitable workplaces, for those of us who make a point to express appreciation for essential workers, and indeed for all who claim that "Black Lives Matter." Highly visible anti-Black violence, such as the murders of George Floyd and Breonna Taylor, is disproportionately placing additional stresses on Black Americans. Our participants' quotes highlight Black Americans' recognition (a) that these types of events are likely to impact them or their family members and (b) that they will have to actively manage their well-being and mental health in light of this knowledge. For some, participation in protests is empowering; for others even watching the news coverage of racist violence and protests against it takes too much of an emotional toll. Practitioners should be aware of this context and recognize that experiences of the dual pandemics will impact their clients differently based on many sociodemographic variables. Although White Americans and other non-Black Americans may be outraged by police violence and feel deep empathy for its victims, they do still have the privilege of choosing whether or not to engage with these issues. Indeed, our participants noted instances in which their White coworkers were clearly not interested in talking about their Black colleagues' experiences and emotional responses related to the dual pandemics and took steps to thwart such conversations. These realities point out the difficult work that needs to be done in many workplaces around the country. Mental health practitioners should be attuned to and actively ask about the workplace experiences of their clients of color, both as those experiences relate to COVID-19 and as COVID-19 wanes. Practitioners should also work to ensure that their work environments are inclusive, equitable, and sensitive to the experiences of individuals from different sociodemographic backgrounds, with particular attention to marginalized communities.

Implications for policy

The results of this study should encourage further examination of our federal policies (e.g., policies concerning essential workers) and how well they center the challenges facing Black American women and other women of color. Clearly, the intersecting identities of race, gender, and essential work status require additional consideration by policymakers, and essential workers should have access to the full range of supports they need to serve in their critical roles for our country, including but not limited to emergency child care, housing support, and proper protective equipment (Frye, 2020). Support may also include providing adequate opportunities for employment and training programs that enhance essential workers' skillsets in order to maximize their earning potential (Frye, 2020). At minimum, policymakers should develop policies and programs that would allow essential workers to be paid a liveable wage. Further, many challenges affecting Black essential workers stem from a history of systemic racism, further revealing the need for interventions to address pervasive social inequality and to offer increased protection for those who risk their lives to protect others (Rodgers et al., 2020).

Conclusion

The lived experiences of Black women essential workers during the COVID-19 pandemic provide researchers, policymakers, and activists with critical insights into the nature of our communities and workplaces. Certainty, at the beginning of the COVID-19 pandemic, many Americans expressed appreciation for essential workers. Our findings, however, point out that sufficient appreciation requires more than slogans, hashtags, and yard signs. Indeed, our participants acknowledged several challenges within and beyond their workplaces. As such, our findings invite further examination of how we as a society care for those who are most immediately "essential" to the functioning of our nation. To truly appreciate essential workers, particularly those who have intersecting marginalized identities, is to work to create a more just and equitable society.

Acknowledgments

The principal investigators would like to thank the research participants who gave of their time to participate in this study.

Availability of Data

The data that support the findings of this study are available on request from the corresponding author, [RG]. The data are not publicly available due to [restrictions e.g., their containing information that could compromise the privacy of research participants].

Disclosure statement

No potential conflict of interest was reported by the author(s).

Funding

This work was supported by the North Carolina Collaboratory, University of North Carolina at Chapel Hill.

ORCID

Rachel W. Goode (iD) http://orcid.org/0000-0002-1358-3917

References

Anderson, E. H., & Spencer, M. H. (2002). Cognitive representations of AIDS: A phenomenological study. *Qualitative Health Research*, *12*(10), 1338–1352. https://doi.org/10.1177/1049732302238747

Babbie, E. (1995). *The practice of social research (7th ed.)* Belmont, CA: Wadsworth Publishing.

Bailey, Z., Barber, S., Robinson, W., Slaughter-Acey, J., Ford, C., & Sealy-Jefferson, S. (2020). *Racism in the time of COVID-19*. Interdisciplinary Association for Population Health Science. iaphs.org/racism-in-the-time-of-covid-19/

Bailey, Z. D., & Moon, J. R. (2020). Racism and the political economy of COVID-19: Will we continue to resurrect the past?. *Journal of Health Politics, Policy and Law*, *45*(6), 937–950. https://doi.org/10.1215/03616878-8641481

Boyatzis, R. E. (1998). *Transforming qualitative information: Thematic analysis and code development*. Sage Publications.

Chandler, R., Guillaume, D., Parker, A. G., Mack, A., Hamilton, J., Dorsey, J., & Hernandez, N. D. (2021). The impact of COVID-19 among Black women: Evaluating perspectives and sources of information. *Ethnicity & Health*, *26*(1), 80–93. https://doi.org/10.1080/13557858.2020.1841120

Frye, J. (2020). *On the frontlines at work and at home: The disproportionate economic effects of the coronavirus pandemic on women of color*. Center for American Progress. https://www.americanprogress.org/issues/women/reports/2020/04/23/483846/frontlines-work-home

Gould, E., & Wilson, V. (2020). *Black workers face two of the most lethal preexisting conditions for coronavirus-racism and economic inequality*. Economic Policy Institute. https://www.epi.org/publication/black-workers-covid

Greig, A., & Taylor, J. (1999). *Doing research with children*. Sage.

Groenewald, T. (2004). A phenomenological research design illustrated. *International Journal of Qualitative Methods*, *3*(1), 42–55. https://doi.org/10.1177/160940690400300104

Guest, G., MacQueen, K. M., & Namey, E. E. (2012). *Applied thematic analysis*. Sage Publications.

Gur, R. E., White, L. K., Waller, R., Barzilay, R., Moore, T. M., Kornfield, S., Njoroge, W. F. M., Duncan, A. F., Chaiyachati, B. H., Parish-Morris, J., Maayan, L., Himes, M. M., Laney, N., Simonette, K., Riis, V., & Elovitz, M. A. (2020). The disproportionate burden of the COVID-19 pandemic among pregnant Black Women. *Psychiatry Research*, *293*, 113475. https://doi.org/10.1016/j.psychres.2020.113475

Kallick, D. D. (2020). *New York's essential workers: Overlooked, underpaid, and indispensable*. Fiscal Policy Institute. https://fiscalpolicy.org/wp-content/uploads/2020/04/Essential-Workers-Brief-and-Recs.pdf

Kantamneni, N. (2020). The impact of the COVID-19 pandemic on marginalized populations in the United States: A research agenda. *Journal of Vocational Behavior*, *119*, 103439. https://doi.org/10.1016/j.jvb.2020.103439

Kinder, M., & Ford, T. N. (2020). *Black essential workers' lives matter. They deserve real change, not just lip service*. The Brookings Institution. https://www.brookings.edu/research/black-essential-workers-lives-matter-they-deserve-real-change-not-just-lip-service/

Kruger, D. (1988). *An introduction to phenomenological psychology (2nd ed.)*. Juta.

National Conference of State Legislators. (2021). *COVID-19: Essential workers in the states*. https://www.ncsl.org/research/labor-and-employment/covid-19-essential-workers-in-thestates.aspx#:~:text=According%20to%20the%20U.,to%20defense%20to%20agriculture

Obinna, D. N. (2020). Essential and undervalued: Health disparities of African American women in the COVID-19 era. *Ethnicity & Health*, *26*(1), 68–79. https://doi.org/10.1080/13557858.2020.1843604

Padgett, D. K. (2017). *Qualitative methods in social work research* (3rd ed.). SAGE Publications, Inc.

Rodgers, T. N., Rogers, C. R., VanSant-Webb, E., Gu, L. Y., Yan, B., & Fares, Q. (2020). Racial disparities in Covid-19 mortality among essential workers in the United States. *World Medical & Health Policy*, *12*(3), 311–327. https://doi.org/10.1002/wmh3.358

Schwandt, T. A. (1997). *Qualitative inquiry: A dictionary of terms*. Sage.

Silverstein, J. (2021). *The global impact of George Floyd: How black lives matter protests shaped movements around the world*. CBS News. https://www.cbsnews.com/news/george-floyd-black-lives-matter-impact/

Snowden, L. R., & Graaf, G. (2021). COVID-19, social determinants past, present, and future, and African Americans' health. *Journal of Racial and Ethnic Health Disparities*, *8*(1), 12–20. https://doi.org/10.1007/s40615-020-00923-3

Thomas, T., Gabbatt, A., & Barr, C. (2020). *Nearly 1,000 instances of police brutality recorded in US anti-racism protests*. The Guardian. October 29. https://www.theguardian.com/us-news/2020/oct/29/us-police-brutality-protest

Tomer, A., & Kane, J. (2020). *To protect frontline workers during and after COVID-19, we must define who they are*. The Brookings Institute. https://www.brookings.edu/research/to-protect-frontline-workers-during-and-after-covid-19-we-must-define-who-they-are/

U.S. Department of Homeland Security. (2020). *Guidance on the Essential Critical Infrastructure Workforce: Ensuring Community and National Resilience in COVID-19 response.* 4.0. Washington, DC: Cybersecurity & Infrastructure Security Agency. https://www.cisa.gov/sites/default/files/publications/ECIW_4.0_Guidance_on_EssentialCritical_ Infrastructure_Workers_Final3_508_0.pdf

Welman, J. C., & Kruger, S. J. (1999). *Research methodology for the business and administrative sciences.* International Thompson.

Demanding migrant/immigrant labor in the coronavirus crisis: critical perspectives for social work practice

Odessa Gonzalez Benson, Fernanda Cross and Christopher Sanjurjo Montalvo

ABSTRACT

The coronavirus pandemic of 2020 laid bare how migrant and immigrant workers are "essential workers" in the critical industries of agriculture/farming, meat production, restaurants/hospitality and health care in the United States. In this article, we discuss this demand for migrant labor and implications for social work. We argue that a labor-focused framework as critical perspective would complement the rights-based, participatory frameworks that inform social work scholarship and practice with immigrants, together accounting for systemic racism, global and national inequality, and discrimination embedded in immigration and social policies and forms of practice. In the first place, by recognizing how non-immigrants and immigrants are inextricably linked through structural means of production and consumption, social workers would develop deeper empathy toward immigrant clients and communities, leading to interactions that are empowering and affirming, and thus effective. Direct practice interventions would be richly informed, as practitioners account for immigrants' work environment, such as difficult work conditions, low wages and lack of benefits, that often impact clients and families. A labor-focused perspective also points to areas of social work advocacy and meso/macro practice, those focusing on workers' rights and immigration policy.

Introduction

Immigrant/migrant[1] labor is the underbelly of the U.S. economy. And as the world turned upside down during the coronavirus pandemic of 2020, this underbelly surfaced. Illustrating what is essential and important in a time of crisis and lockdown, the pandemic revealed the value of and demand for the laboring of im/migrants. In this article, we consider im/migrant labor in the context of the coronavirus pandemic, examining vulnerabilities faced by im/migrants and their families, and implications for social workers. First, we describe im/migrant labor in essential industries: agriculture/farming, meat production, restaurants/hospitality, health care. Second, we illustrate the precarity embedded in these industries, rendered all the more acute by the pandemic. Third, we discuss demands for racialized im/migrant labor, laid bare by the coronavirus crisis and border closure. We close by calling for critical, structural perspectives on social work practice and our understanding of im/migrants' place in our local and national community, and offer suggestions for steps moving forward.

Im/migrant labor in essential industries

Revenue from the agriculture/farm industry contributes to a significant portion of the U.S. economy. Mexican-origin migrants comprise the largest percentage of farmers, making up approximately 57% of all farm laborers, graders, and sorters (U.S. Department of Agriculture, 2020). California, which produces over a third of the country's vegetables and two-thirds of the country's fruit and nuts, relies overwhelmingly on migrant labor, comprising 70% of the state's farmworkers (Neuburger, 2019). The meatpacking industry relies on immigrants, who make up about 40% of all meatpacking workers (Migration Policy Institute, 2020). In 2000, Latino/a workers held 82% of the meat-processing jobs, engaged in labors such as receiving and killing, evisceration and inspection, cutting and deboning, processing and packing, and sanitation and cleaning (Kandel, 2006). Restaurants employ more than 2 million immigrants in essential roles such as cooks, dishwashers, and food preparation (Shierholz, 2014). In hospitality/tourism, one in five workers was an immigrant in 2018, working as housekeepers, janitors, front desk clerks, and cleaners (New American Economy, 2020). In health care, one in six workers (i.e., doctors, nurses, home health aides) nationally is an immigrant (Asian, 42%; Latin American, 18%; Caribbean, 16%); while in some states, such as New York and New Jersey, immigrants account for over half of all low-wage health workers such as home health aides and nursing assistants (Altorjai & Batalova, 2017).

Risks and precarities in Im/migrant industries heightened during pandemic

Migrants and particularly undocumented migrants are essential workers facing risks and precarities, as are immigrants with residency and legal status, who, despite their immigration status, experience hardships intergenerationally. Many im/migrant workers have no health care benefits, paid sick leave, or unemployment insurance – protections particularly crucial during the pandemic (Cross and Gonzalez Benson, 2021). Farmworkers feel more vulnerable than ever as they work without the most basic protective equipment, such as masks and soap (Shoichet, 2020). They have a higher risk of getting sick and spreading the virus to others due to poor work and living environments, often sharing overcrowded quarters with others. Fearing deportation or job loss, workers rarely report substandard work environments (Willingham & Mathema, 2020). The COVID-19 pandemic made Latino/a immigrants' worse overall health and their lack of available health care starkly evident (Bosman et al., 2021; Cross and Gonzalez Benson, 2021). Also, many immigrant workers in precarious jobs face numerous challenges when seeking to claim their workers' rights and are often not informed of such rights to begin with.

The virus ran rampant in the meatpacking industry, infecting thousands of workers in only a matter of weeks. Many im/migrant workers were pressured to continue working in enclosed spaces in plants, without adequate protective equipment or social distancing measures. Some employers offered cash bonuses to workers who did not miss any of their shifts, incentivizing them to show up to work even while sick. As a result, several meat plants were forced to close or drastically reduce production (Runyon, 2020). In an affronting paradox, despite being frontline workers, migrants were excluded from the government's stimulus plan. Many essential migrant workers were thus left to fend for themselves as they faced financial uncertainty (Jarvie, 2020; Cross and Gonzalez Benson 2021).

The coronavirus pandemic reveals our demand for Im/migrant labor

While an unprecedented number of workers were laid off or furloughed due to the pandemic, many im/migrant workers constituted a workforce deemed vital in keeping the country going. Not only in the United States but globally, as national borders closed off the world with travel restrictions to protect citizens from the virus, im/migrant workers alone were exempted from travel bans.

In Canada, while everyone else was banned, seasonal farmworkers were allowed entry, their work deemed "absolutely critical ... to maintain the food security of all Canadians" (Bensadoun, 2020). Similarly, in the United Kingdom, fruit pickers from Romania were flown into the country in flights chartered by employers, despite travel restrictions (O'Carroll, 2020). In a nationwide recruitment campaign called Feed the Nation, the UK government called upon students and furloughed employees to work on farms. However, recruitment failed; 90,000 farmworkers were needed but only 35,000 applied. A mere 5,500 came for interviews, and only 1,650 of interviewees had the needed experience and skills to do the job (O'Carroll, 2020).

In the United States, agricultural employers gave out letters noting that the Department of Homeland Security deemed migrant workers "critical to the food supply chain" (Jordan, 2020). In health care, special work visas were created for health care professionals from other countries, such as nurses from the Philippines so they could work in overwhelmed U.S. hospitals serving patients with COVID-19 (Elemia, 2020). Meanwhile, in meatpacking industries, an executive order prevented companies from temporarily closing, citing them as critical infrastructure, without regard for health and safety (Nareaa, 2020).

Im/migrant labor: critical perspectives in social work practice

Social work theories on immigration often rely on notions of rights to citizenship and membership in places of migration (Bloemraad, 2018; McPherson, 2020; Valtonen, 2001). Im/migrants are presented as rights-bearing individuals (McPherson, 2020) and/or as contributing residents – as business owners, taxpayers, and members of a cultural community – yielding social, cultural, and economic benefits for all by forging a more diverse national and local community (Valtonen, 2001). These frameworks are based upon the idea that im/migrants have the right to move as they seek better opportunities and/or escape poverty and danger in home countries. A central premise is the salience of push factors, those individual-level factors internal to im/migrants themselves and environmental factors in their home countries that compel migrants to move.

Intersecting with those rights-based, participation-based perspectives on immigration, there are pull factors in countries of im/migration: the demand for migrant labor. We argue that incorporating im/migrant labor into conceptual understanding of immigration would promote social justice perspectives and strengthen arguments for immigrant rights. A rights-based, participation-based, *and* labor-focused perspective would encompass the systemic racism, global-national inequality and discriminatory immigration policing and detention, and their adverse, painful impacts upon immigrant communities.

Social work practice, education, and research would be enriched with this labor-focused framework, by integrating it or using it concurrently with rights-based and participatory-based frameworks on immigration. We have illustrated the following three points in the three sections above. A first step is for social workers to move discussions about individual level push factors to also include structural factors – labor demand – and our complicity in it as consumers. Second, social workers can share arguments, with data and factual information both historical and current (such as during the pandemic), about the centrality of im/migrant workers in U.S. industries. Third, social workers can make evident the risks, precarities, and exploitation that such jobs involve, and how they are linked with racial-ethnic divides and structural racism that advantage some while disadvantaging and exploiting the immigrant "other." Exclusions and exploitations in im/migrant labor stem from ethnic and racial inequalities, resulting in limited job opportunities, pay inequity, and hazardous and poor working conditions, among other job-related outcomes. A labor-focused lens allows a structural analysis of immigration that takes account of systemic, institutional, and ideological means of exploitation and benefit from im/migrant labor, rather than a person-centered analysis focused on im/migrants' employment and personal pursuits.

Using critical perspectives on immigration that take account of our social demand for im/migrant labor has further implications for social work. First, by recognizing how non-immigrants and immigrants are inextricably linked as consumers and producers, social workers would acquire deeper social empathy, "the ability to more deeply understand people by perceiving or experiencing their life situations and as a result gain insight into structural inequalities and disparities" (Segal, 2011, p. 266; Cross & Gonzalez Benson, 2020). Greater empathy itself would lead to reflection about social responsibility. Second, employing a critical, labor-focused lens on immigrants calls upon direct practitioners and those who conduct interventions to consider im/migrants' work environment, such as difficult work conditions, low wages, and lack of benefits, that impact clients' and families' well-being and success (Ayón, 2014). Third, this lens also calls for participatory approaches to community development and community building work at the meso-level of social work, and for collaborating with immigrants themselves and their organizations toward robust integration (Gonzalez Benson, 2020). A labor-focused perspective on immigration would also point to social work advocacy, practice, and research that focuses on workers' rights and immigration policy (Cleaveland, 2010).

The coronavirus pandemic – in revealing our demand for im/migrant labor – inadvertently presents us with an opportunity to adopt a more nuanced, critical perspective on immigration. Immigration is about im/migrants' rights, participation, needs and desire to move, but it is also about our demand for im/migrant labor. We as consumers are thus implicated in immigration, while the social work profession offers means for directly applying critical, structural, labor-focused perspectives toward transformative practice with im/migrants.

Note

1. This article uses the term "im/migrant" to denote both "migrant" and "immigrant" and refers to each term specifically as applicable. Conventionally, a migrant is someone who is temporarily in a host country for work or other reasons (i.e., seasonal farm workers, IT temp workers at Google, visiting scholars), while an immigrant is someone who was/is pursuing a legal process of adjusting status as permanent resident (i.e., first generation). However, these distinctions are dynamic and nuanced, rather than static legal definitions. Some migrants may be in immigration processes or may intend/aim to immigrate in the future, while some immigrants (i.e., undocumented) may be residing or intend to reside permanently without means for legal immigration processes. Definitions aside, the labor-focused perspective we discuss in this article applies to both migrants and immigrants, in varying aspects.

Disclosure statement

No potential conflict of interest was reported by the author(s).

References

Altorjai, S., & Batalova, J. (2017, June 28). *Immigrant health-care workers in the United States*. Migration Policy Institute. https://www.migrationpolicy.org

Ayón, C. (2014). Service needs among Latino immigrant families: Implications for social work practice. *Social Work*, *59* (1), 13–23. https://doi.org/10.1093/sw/swt031

Bensadoun, E. (2020, April 5). Canada to allow seasonal foreign workers but they must self-isolate, minister says. *Global News*. https://globalnews.ca/news/6780779/seasonal-farm-workers-coronavirus/

Bloemraad, I. (2018). Theorising the power of citizenship as claims-making. *Journal of Ethnic and Migration Studies*, *44* (1), 4–26. https://doi.org/10.1080/1369183X.2018.1396108

Bosman, J., Kasakove, S., & Victor, D. (2021, July 21). U.S. life expectancy plunged in 2020, especially for Black and Hispanic Americans. *New York Times*. https://www.nytimes.com/2021/07/21/us/american-life-expectancy-report. html

Cleaveland, C. (2010). "We are not criminals": Social work advocacy and unauthorized migrants. *Social Work*, *55*(1), 74–81. https://doi.org/10.1093/sw/55.1.74

Cross, F. L., & Gonzalez Benson, O. (2021). The Coronavirus pandemic and immigrant communities: A crisis that demands more of the social work profession. *Affilia: Journal of Women and Social Work*, *36*(1), 113–119. https://doi. org/10.1177/0886109920960832

Elemia, C. (2020, March 21). Germany to fly in Filipino nurses to care for their coronavirus patients—report. *Rappler*. https://amp.rappler.com/nation/255388-germany-hesse-filipino-nurses-coronavirus-patients-covid-19

Gonzalez Benson, O. (2020). Welfare support activities of grassroots refugee-run community organizations: A reframing. *Journal of Community Practice*, *28*(1), 1–17. https://doi.org/10.1080/10705422.2020.1716427

Jarvie, J. (2020, April 20). These U.S. citizens won't get coronavirus stimulus checks—because their spouses are immigrants. *Los Angeles Times*. https://www.latimes.com/world-nation/story/2020-04-20/u-s-citizens-coronavirus-stimulus-checks-spouses-immigrants

Jordan, M. (2020, April 2). Farmworkers, mostly undocumented, become 'essential' during pandemic. *New York Times*. https://www.nytimes.com/2020/04/02/us/coronavirus-undocumented-immigrant-farmworkers-agriculture.html

Kandel, W. (2006, June 1). *Meat-processing firms attract Hispanic workers to rural America*. U.S. Department of Agriculture Economic Research Service. https://www.ers.usda.gov/amber-waves/2006/june/meat-processing-firms-attract-hispanic-workers-to-rural-america/

McPherson, J. (2020). Now is the time for a rights-based approach to social work practice. *Journal of Human Rights and Social Work, 5*(2), 61–63. https://doi.org/10.1007/s41134-020-00125-1

Migration Policy Institute. (2020, April 14). *The essential role of immigrants in the U.S. food supply chain*. https://www.migrationpolicy.org/content/essential-role-immigrants-us-food-supply-chain

Nareaa, N. (2020, April 30). Trump is keeping meatpacking plants open—but employees are scared to show up for work. *Vox*. https://www.vox.com/2020/4/30/21241167/meatpacking-workers-coronavirus-tyson-smithfield

Neuburger, B. (2019, June 1). California's migrant farmworkers. *Monthly Review*. https://monthlyreview.org/2019/05/01/californias-migrant-farmworkers/

New American Economy. (2020, April 22). *Hospitality & tourism*. https://www.newamericaneconomy.org/issues/hospitality-&-tourism/

O'Carroll, L. (2020, April 17). British workers reject fruit-picking jobs as Romanians flown in. *The Guardian*. https://www.theguardian.com/environment/2020/apr/17/british-workers-reject-fruit-picking-jobs-as-romanians-flown-in-coronavirus

Runyon, L. (2020, April 19). *Meatpacking plant working conditions stoke spread of coronavirus*. National Public Radio. https://www.npr.org/2020/04/19/838195049/meatpacking-plant-working-conditions-stoke-coronavirus-spread

Segal, E. (2011). Social empathy: A model built on empathy, contextual understanding, and social responsibility that promotes social justice. *Journal of Social Service Research, 37*(3), 266–277. https://doi.org/10.1080/01488376.2011.564040

Shierholz, H. (2014, August 21). *Low wages and few benefits mean many restaurant workers can't make ends meet*. Economic Policy Institute. https://www.epi.org/publication/restaurant-workers/

Shoichet, C. (2020, April 11). The farmworkers putting food on America's tables are facing their own coronavirus crisis. *CNN*. https://www.wsmv.com/news/us_world_news/the-farmworkers-putting-food-on-americas-tables-are-facing-their-own-coronavirus-crisis/image_1ce331b7-cf76-5e5d-8b71-a8406b2481c0.html

U.S. Department of Agriculture. (2020, April 22). *Farm Labor*. https://www.ers.usda.gov/topics/farm-economy/farm-labor/

Valtonen, K. (2001). Social work with immigrants and refugees: Developing a participation-based framework for anti-oppressive practice. *British Journal of Social Work, 31*(6), 955. https://doi.org/10.1093/bjsw/31.6.955

Willingham, Z., & Mathema, S. (2020, April 23). *Protecting farmworkers from coronavirus and securing the food supply*. Center for American Progress. https://www.americanprogress.org/issues/economy/reports/2020/04/23/483488/protecting-farmworkers-coronavirus-securing-food-supply/

Mask mandates, race, and protests of summer 2020

Rahbel Rahman ⓘD, Sameena Azhar ⓘD, Laura J. Wernick ⓘD, Jordan E. DeVylder ⓘD,
Tina Maschi ⓘD, Margaret Cohen and Simone Hopwood

ABSTRACT
This study examined predictors to mask mandate support and racial justice
protest participation across Asian (n = 103), Black (n = 102), white (n = 102)
New York City residents, using binary logistic regressions. Participants with
positive feelings about the racial justice movement were more likely to participate in the protests. White and Asian respondents were more likely to support
the mask mandates over Black respondents. Asian respondents were less likely to
participate in public protests over white respondents. Our findings offer a model
for social workers to understand how race, political participation and COVID-19
intersect to create racially just responses to health and justice matters.

Introduction

COVID-19 and police brutality are two intersecting calamities (Jean, June 16, 2020). Over the summer
of 2020 and in the midst of the COVID-19 pandemic, Americans witnessed the extrajudicial killings of
Ahmaud Arbery, Breonna Taylor, George Floyd, and several other Black people at the hands of police
officers (Gibson et al., 2020). The COVID-19 pandemic has brought social and racial injustice and
inequity to the forefront of public health as Black, Indigenous and other People of Color (BIPOC) have
disproportionately been infected by and died of COVID-19. These dual pandemics have ignited
a sense of fervor and rage as people acknowledge the devastating impacts of racism, systemic
oppression, and police brutality (Gibson et al., 2020).

There have been 10,600 demonstrations across the United States between May 24 and August 22,
2020, shortly following the killing of George Floyd in police custody in May 2020, of which 7,750 (80%)
were linked to either the Black Lives Matter (BLM) movement or the COVID-19 pandemic (Kishi &
Jones, 2020). These protests took place in more than 2,440 locations across all 50 states making these
protests the largest social movement in the country's history (Kishi & Jones, 2020). The BLM movement,
which was founded in response to the killing of Trayvon Martin in 2013, has used protests, social media,
and publicity to shed light on racial discrepancies in treatment by the police. The BLM movement has
also underscored the implications of systemic racism on communities of color (Black Lives Matter,
2021). Within the context of these dual pandemics, this study sought to examine predictors to mask
mandate support and racial justice protest participation across 308 residents of New York City (NYC)
who identified as Asian (n = 103); Black (n = 102) and white (n = 102).

COVID-19 and New York state response

NYC was the epicenter of the COVID-19 pandemic during the first wave of infections in the United
States, roughly from March to June, 2020. The spread of COVID-19 has been attributed to respiratory
droplets, emitted through talking, laughing, coughing, breathing, or sneezing (Centers for Disease

Control and Prevention [CDC], 2021a). In an effort to reduce the spread of COVID-19, public health experts urged the public to avoid large social gatherings in public settings, including protests; to maintain six feet distance apart from others outside of their household; and to engage in public mask-wearing and frequent hand-washing (Centers for Disease Control and Prevention [CDC], 2021b).

Racial disparities in health and mortality are long-standing and consistently replicated over time, through evolving mechanisms (Phelan & Link, 2015); two highly relevant factors in recent history are police killings and COVID-19. Along with a higher incidence of police-inflicted killings (Edwards et al., 2019), Black communities have also experienced disproportionately higher rates of COVID-19 infections, hospitalizations, and death over other racial groups (Centers for Disease Control and Prevention [CDC], 2021c). Black people are 2.9 times more likely to experience hospitalization and 1.9 times more likely to die from COVID-19 than white people (Centers for Disease Control and Prevention [CDC], 2021d). Similarly, Hispanic/Latinx people are 3.1 times more likely to be hospitalized and 2.3 times more likely to die from COVID-19 than white people (Centers for Disease Control and Prevention [CDC], 2021d).

Within the context of COVID-19, racialized bodies continue to be systematically subjected to racism, exploitation, and erasure. Approximately 75% of workers in the United States could not perform their job at home through the pandemic; these jobs tended to be disproportionately filled by BIPOC who are considered "frontline or essential workers" (Krieger, 2020). Some of those who were employed as essential workers sacrificed their lives to jump start the economy (Gibson et al., 2020). At the beginning of the pandemic, many essential workers did not have access to personal protective equipment (PPE), further placing them at risk for COVID-19 infection (Gibson et al., 2020).

To contain the spread of COVID in New York State, Governor Cuomo declared a State of Emergency by announcing a series of actions, including mask mandates and issuing quarantine periods for travelers arriving outside of the tristate area of New York, New Jersey and Connecticut (New York State [NYS] Governor's Office, 2020a). Through an executive order in March 2020, Cuomo declared New York State to be "on pause," effectively closing all non-essential businesses and initiating shelter-in-place orders. This stipulated guidance for the termination of large public gatherings; the temporary suspension of instruction in public schools; and restrictions on indoor gatherings at restaurants, bars, and public spaces (New York State [NYS] Governor's Office, 2020b). In May and June of 2020, the governor also imposed a mandatory curfew, mandating all New Yorkers, including protestors, to wear face coverings in public (New York State [NYS] Governor's Office, 2020c). However, previous state mask laws were in direct conflict with these public health orders. New York Penal Law 240.35(4), which was in place since 1845, prohibited two or more people wearing masks or any face covering from congregating in a public place; violators were subject to 15 days in prison (Mahbubani, 2020). The law was only repealed in May 2021 (Mahbubani, 2020).

Mask mandates and mask-wearing

Since early in the pandemic, there has been scientific agreement regarding the use of facial coverings to slow, if not prevent, the spread of COVID-19 infection (Martinelli et al., 2021). By mid-July 2020, masks were mandatory in 21 states, with additional states considering the adoption of such policies (Gostin et al., 2020). Yet mask usage in the United States has proved controversial across racial, cultural, and political allegiances, and has only been selectively adopted.

While primarily a protective measure, the COVID-19 mask has also become a cultural icon. For some, it has become a marker of social responsibility and good citizenship (Martinelli et al., 2021). It can represent the wearer's compliance with public safety and communal well-being through exercising care for one's self and others. Some individuals have been unwilling to wear masks, either calling the virus a "hoax," attributing the virus to the Chinese government, supporting statements that COVID-19 was no more dangerous than the seasonal flu, or believing that masks were ineffective or unnecessary for prevention (Taylor & Asmundson, 2021). Some white Americans refused to wear masks, citing mask mandates as a violation of their civil liberties though remaining relatively unafraid of facing police action for their violations (Lawrence, 2020).

Conversely, many Black Americans have experienced a paradox during the pandemic: (1) wear a mask and risk police interaction due to racial profiling or (2) opt not to wear a mask and risk police interaction for violating mask mandates (Lawrence, 2020). Within New York, Black people made up 93% of coronavirus-related arrests (Taylor, 2020). Black and Hispanic/Latinx Americans may not endorse mask mandates as a result of a legacy of policing Black and Brown bodies due to their clothing and appearance (Lowe, 2020). Moreover, legacies of medical and research abuse, including the infamous Tuskegee syphilis study, have left some Black communities weary of trusting government or medical sources. While racial/ethnic minority communities have increased morbidity and mortality from COVID-19, there is sometimes heightened reluctance in these communities to get vaccinated (Kaiser Family Foundation [KFF], 2021). Medical mistrust is therefore another source of contention for supporting mask mandates.

Black Lives Matter movement and racial justice protests of summer 2020

Black Lives Matter (BLM) has become a widespread social movement to undo systemic racism. The hashtag #BlackLivesMatter was introduced in 2013 in response to the acquittal of Trayvon Martin's murderer; it went viral on social media. Contemporary law enforcement activity remains heightened for BIPOC and is hinged to a legacy of racial superiority (Lawrence, 2020). Historical and current policing of Black communities relies on the presumption of Black hostility or the belief that Black individuals are less law-abiding (Lowe, 2020). Black men are 2.5 times more likely to be killed by police over their life course than white men and women; similarly, Black women are about 1.4 times more likely to be killed by police (Edwards et al., 2019). Punitive criminal justice policies, such as the War on Drugs and the Violent Crime and Law Enforcement Act, have targeted Black and Hispanic/Latinx communities, also contributing to mass incarceration (Lowe, 2020). Consequently, far more Black men and women are imprisoned today than were enslaved prior to the Civil War, contributing to a longstanding mistrust of Black communities toward the police (Lowe, 2020).

In the United States, bystanders recording death events on their cell phones and police webcams capturing incidents of police brutality, have led to the recreation of lynching as a public spectacle (Dreyer et al., 2020). Some white individuals have felt a combination of shame, guilt, and anger and hence mobilized themselves to question the status quo of white supremacy (Dreyer et al., 2020). The BLM movement has gained new supporters who are willing to put themselves at risk and protest racism and police brutality during a pandemic. The racial justice protests of summer 2020 raised questions regarding the moral compass of the United States, particularly by younger people. The protests were not only calling for an end to injustice, police brutality, race-based violence, and institutional racism, but also for an awareness of these problems in socio-economic and political systems that entrench structural power and privilege and then place the onus on victims for their own social oppression. To engage in protest during a pandemic highlighted how, like COVID-19, the pandemic of racism also kills a larger number of people daily (Çetinkaya, 2020). Widespread public engagement in these protests reflects a deep commitment to address racism, even in the midst of public health threats to safety.

In more than 94% of the racial involved justice protests, demonstrators have engaged in peaceful protest; only a minority of protests reported violence, clashes with police, vandalism, looting, or destructive activity (Kishi & Jones, 2021). It is unclear whether the violence was provoked by demonstrators or due to aggressive government action, intervention from right-wing groups or individual assailants, and car-ramming attacks (Kishi & Jones, 2021). However, according to polls, a substantial proportion of people watching the protests at home perceived them to be violent (Cage, 2020). A possible explanation for this is the impacts of biased media framing or the language used by Trump to describe participants of the racial justice protests as "thugs," "mobs," and engaging in "acts of domestic terror" (Cage, 2020). Alternatively, demonstrations involving right-wing militias or militant social movements, which often opposed the lock-down measures and mask-wearing, turned violent or destructive nearly 14% of the time (Kishi & Jones, 2021).

CDC's advisory to wear masks came almost two months prior to the summer 2020 racial justice protests. However, due to the risk of disease transmission, public health experts and policy makers were at a juxtaposition of accommodating the constitutional right to freely assemble and the need to control community risk through the prohibition of large-scale events (Kampmark, 2020). While most organizers asked protesters to wear masks and social distance, critics of the racial protests have implied that participation in street protests meant that one was simultaneously being irresponsible in spreading the virus and contributing to the undermining of the lockdown's credibility (Kampmark, 2020). Despite the anticipated risks of increased infections during public gatherings, the increase in COVID-19 infections through an individual protester has not been found to be statistically significant (Lazer et al., 2021). In light of these various issues impacting the dual pandemics, we sought to conduct an exploratory study of factors predicting attitudes regarding mask mandates and participation in racial justice protests.

Theoretical framework

We applied Critical Race Theory (CRT) in this study in an effort to center racism in our understanding and meaning-making of the racialization of different group attitudes regarding mask mandates and political participation. BLM has arguably adopted the ideals of CRT, which focus on reforming a systemically racist society (Dixson, 2018). CRT scholars contend that race is a social construct that permeates American society and that historically created social policies that continue to create advantages for whites and marginalize BIPOC (Delgado & Stefancic, 2017). Ideologies regarding white supremacy help enable the murder of George Floyd and countless other Black men with the relative dismissal from punishment for the officers involved in these killings. Applying CRT, our paper sought to understand racialized perceptions regarding mask mandates and political protests during a dual pandemic.

Methods

Sampling and recruitment

We conducted an online study of 308 New Yorkers on their attitudes regarding mask mandates and the summer 2020 racial justice protests. The eligibility criteria for the study were that participants be current residents of NYC and over the age of 18 years. Their residency was verified using self-reported questions regarding borough of residence and zip code. Zip codes were verified by Qualtrics as being within NYC limits. Eligibility was further qualified through stratified sampling, resulting in roughly equal numbers of Asian (n = 103), Black (n = 103) and white (n = 102) participants. Race was determined by a self-report question; respondents were asked to respond to a multiple-choice question with options following the U.S. Census Bureau's categories of race: American Indian/Alaska Native, Asian, Black/African American, Native Hawaiian/Pacific Islander, and white. Participants were also able to indicate if they identified as biracial or multiracial.

Data collection

Data collection was completed by Qualtrics XM, an online survey platform. Qualtrics set the incentive amount for participants at a range from $2.50 to $4. The study protocol was approved by an Institutional Review Board (IRB). Data was collected between August to November 2020.

Measures

Dependent variables

Protest participation was measured by whether or not the respondent had participated in protests or gatherings related to racial justice issues during the summer of 2020. *Mask mandate support* was measured by whether participants were in favor of (positive association) or against (negative association) the mask mandates.

Predictors

COVID diagnosis/symptoms. Because COVID-19 tests were not readily available and/or accessible prior to this study, the two measures of having been diagnosed with COVID-19 and having had symptoms of COVID-19 were combined into one binary variable, indicating whether or not participants had been diagnosed with or had symptoms of COVID-19. *Perceptions about the racial justice movement* was measured by whether participants had positive or negative feelings about the racial justice movement. Regarding demographics, *Age* was measured in years. *Ethnicity* included Hispanic/Latinx or not Hispanic/Latinx. *Race* categories included Asian, Black, white. *Gender* was categorized as cisgender man, cisgender woman, and Transgender or Gender Non-Conforming (TGNC). *Sexual orientation* was categorized as heterosexual and not heterosexual.

Data analysis

Descriptive statistics were performed to characterize the overall sample and for each racial group, namely Asians, Blacks and whites. Percentages were reported for categorical variables. Means and standard deviations (SD) were reported for continuous variables. Chi-squared tests were run on the descriptive statistics to examine statistical differences across racial groups and the full sample. Binary logistic regressions were performed to examine group differences among Black, Asian, and whites' attitudes toward mask mandates support and participation in racial justice protests. Hosmer-Lemeshow test was used to select the most parsimonious model and to evaluate the goodness of fit. Data were presented as adjusted odds ratios (aORs) with 95% confidence intervals (CIs) with p-value of <0.05 deemed as significant. Some variables that did not indicate a statistically significant bivariate correlation with protest participation, including sexual orientation and Hispanic/Latinx ethnicity, were removed from the final models. Calculations were completed using STATA Statistical Software (Release 14).

Results

Detailed sociodemographic characteristics of the total sample and by race are provided in Table 1. Within the total sample of 308 respondents from NYC, 103 (33.4%) identified as Black, 103 (33.4%) as Asian, and 102 (33.1%) as white. Eighty-eight percent of the sample identified as heterosexual (n = 271). The majority of the sample identified as cisgender women (n = 181; 59%), followed by cisgender men (n = 119; 38%) and TGNC (n = 6; 2%). Hispanic/Latinx respondents were more likely to identify as Black (n = 29, 28.2%), and less likely to identify as Asian (n = 6, 5.8%), compared to the full sample (n = 59, 19.2%). Among cisgender men, respondents disproportionately identified as white (n = 51, 50%) compared to the full sample (n = 119, 38.8%). Cisgender women were less likely to identify as white (n = 49, 48.5%) compared to the full sample (n = 181, 59%). On average, Black respondents were younger (M = 31.6, SD = 13.85) and whites were older (45.7, SD = 18.8), compared to the full sample (M = 37.3, SD = 16.32).

Sixty-six participants in the study (21.4%) reported they had symptoms and/or had been diagnosed with COVID-19. The majority of the sample (n = 163, 52.9%) indicated that they had participated in protests over the summer of 2020. Black respondents were less likely to support the mask mandate

Table 1. Descriptive statistics for total sample and by race.

Variables	Full sample (n = 617)		Hispanic White (n = 49)		Hispanic Black (n = 55)		NonHispanic Asian (n = 22)	
	n	%	n	%	n	%	n	%
Race								
Black	103	33.44	103	33.44	–	–	–	–
Asian	103	33.44	–	–	–	–	103	33.44
White	102	33.12	–	–	102	33.12	–	–
Ethnicity								
Hispanic/Latinx	59	19.16	**29**	**28.16**	24	23.53	**6**	**5.83**
Non-Hispanic/Latinx	249	80.84	**74**	**71.84**	78	76.47	**97**	**94.17**
Sexual Orientation								
Not-heterosexual	37	12.01	15	14.56	13	12.75	9	8.74
Heterosexual	271	87.99	88	85.44	89	87.29	94	91.26
Gender								
Cisgender man	119	38.76	31	30.10	**51**	**50.00**	37	36,27
Cisgender woman	181	58.96	68	66.02	**49**	**48.51**	64	62.14
TGNC*	6	1.96	4	3.88	1	0.99	1	0.98
Mask Mandate Support								
Yes	211	81.18	**57**	**71.25**	72	85.71	82	86.32
No	48	18.53	**23**	**28.75**	12	14.29	13	13.68
COVID symptoms/ diagnosis								
Yes	66	21.43	23	22.33	27	25.49	17	16.60
No	242	78.51	80	77.67	76	74.51	86	83.50
Feelings about racial justice movement								
Positive	166	59.71	**67**	**73.63**	48	53.93	51	52.04
Negative	112	40.29	**24**	**26.37**	41	46.07	47	47.96
Attended at least one protest								
Yes	163	52.92	60	58.25	56	54.90	47	45.63
No	145	47.08	43	41.75	46	45.10	56	54.37
	M	SD	M	SD	M	SD	M	SD
Age (in years)	37.33	16.32	**31.54**	**13.85**	**45.75**	**18.89**	34.80	12.05

Values in bold indicate categories where significant differences (p < .05) between respondents' race and the full sample (based on the results of χ2 test, respectively, independent-samples median test for age).

(n = 67, 71.3%) compared to the full sample (n = 211, 81.2%). However, Black respondents were more likely to have positive feelings toward the racial justice movement (n = 67, 71.3%), compared to the full sample (n = 166, 59.7%).

Perceptions about the mask mandates

White (aOR = 2.88; 95% CI = 1.10, 7.55) and Asian (aOR = 31.5; 95% CI = 1.34, 7.40) respondents were more likely to support the mask mandates over Black respondents. Those who had positive feelings about the racial justice movement were more likely to support the mask mandates (aOR = 2.68; 95% CI = 1.29, 5.57). Cisgender men were less likely to support the mask mandates (aOR = 0.41; 95% CI = 0.20, 0.84) over cisgender women and TGNC. People who previously had COVID-19 symptoms or a diagnosis (aOR = 0.42; 95%, CI = 0.19, 0.94) were less likely to support the mask mandates over those who did not have symptoms or a diagnosis. The overall model yielded a χ2 (7, 255) = 33.63, $p < .001$. Please refer to Table 2.

Participation in public protests

Age and protest participation were inversely related, such that younger people were more likely to participate in protests (aOR = 0.95; 95%CI = 0.93, 0.97). Cisgender men were more likely (aOR = 2.56, 95% CI = 1.42, 4.61) to participate in protests over cisgender women and TGNC people. Asian respondents were less likely to participate in public protests (aOR = 0.39, 95% CI = 0.18, 0.83) over

Table 2. Binary logistic regression models predicting mask mandate and protest participation.

Variable	Support of Mask Mandate			Protest participation		
	B	OR	95% CI for OR	B	OR	95% CI for OR
Intercept (ref. Black)	0.17	1.18		1.00	2.70	
Intercept (ref White)	1.22	3.40		1.75	5.78	
White (ref. Black)	1.06	2.88	[1.10, 7.55]*	0.75	2.13	[0.95, 4.76]
Asian (ref. Black)	1.15	3.15	[1.34, 7.40]**	−0.03	0.97	[0.49, 1.94]
Asian (ref. White)	0.09	1.10	[[0.42, 2.87]	−0.79	0.46	[0.22, 0.96]*
Age (in years)	0.02	1.03	[1.00, 1.06]	−0.05	0.95	[0.93, 0.97]***
Cisgender man (ref. TGNC & cisgender woman)	−0.90	0.41	[0.20, 0.84]*	0.94	2.56	[1.42, 4.61]**
Support of Mask Mandate				−0.37	0.69	[0.33, 1.45]
Feeling about racial justice movement	0.98	2.68	[1.29, 5.57]**	1.11	3.02	[1.38, 6.63]**
COVID-19 symptoms/diagnosis	−0.87	0.42	[0.19, 0.94]*	0.34	1.41	[0.78, 2.53]
Protest participation	−0.38	0.68	[0.31, 1.48]			
Hosmer-Lemeshow test (sig)	2.50 (0.96)			4.71 (0.79)		
R^2	.14			.16		
χ^2	33.63***			57.10***		

* $p < .05$ ** $p < .01$ *** $p < .001$

ref. = reference category; TGNC = transgender gender nonconforming. Separate models were run using Black and then White as reference groups, in order to test for significant differences between all race categories. This did not affect the other variables in the regression analyses.

white respondents. Those who had positive feelings about the racial justice movement were more likely to participate in the protests (aOR = 3.02, 95% CI = 1.38, 6.63). The overall model resulted in a χ2 (7, 255) = 57.10, $p < .001$. Please refer to Table 2.

Discussion

The summer of 2020 was an intense period in the United States where there were dual dangers to public health, namely the COVID-19 pandemic and incidents of police brutality against the Black community. Our study findings are somewhat consistent with previous research on mask mandates. In a prior study, white (43.8%) participants, followed by Black (43.4%), Asian (31.7%), and then Multiracial (28.3%) participants, indicated that state and law enforcement were doing a good job handling mask mandates (Lawrence, 2020). In our own study, Asian and white respondents were more likely to support mask mandates than Black respondents. For East Asian respondents, there may have been an increased likelihood of supporting the mask mandate as mask-wearing is often perceived to be a collective, civic duty in East Asian countries, for one's own health as well as the safety of others (Wong, 2020). Mask-wearing is widely normalized in East Asian countries and immigrants in the East Asian diaspora communities, including those in the United States, typically follow this trend (Wong, 2020). This may explain their increased support for mask mandates. Contrary to the literature that suggests that some white Americans contend that mask requirements infringe on their civil rights (Taylor & Asmundson, 2021), white respondents in our study were more likely to support mask mandates.

The survey was conducted at a time when COVID-19 testing was limited in NYC. Black respondents who had typical COVID-19 symptoms were more likely to have had the illness, and perhaps developed some level of immunity. They may therefore be less likely to support the mask mandates as they believed they had some form of protection against COVID-19 re-infection. Another plausible explanation for not endorsing mask mandates could be that they did not trust the policy to be equitably applied. The cultural politics behind wearing masks exposes another basis for police interaction and selective enforcement against BIPOC. The concept of "mask tipping," which asks Black individuals to show shopkeepers their faces when entering stores, was used as a means to perpetuate anti-Black racism by requiring racialized bodies to reveal themselves as "safe" to avoid biases and endangerment from store owners (Jan, 2020).

Black respondents may also be hesitant to endorse the mask mandate because of a longstanding history of medical mistrust and deceit from public health entities. A number of unethical incidents, ranging from the 1972 Tuskegee study to the usage of Henrietta Lacks' stems cells without her consent, medical institutions have had histories of misleading minoritized communities, which contributes to current medical mistrust and reduced patient-physician engagement of Black men (Alsan & Wanamaker, 2018). Further qualitative research is warranted to understand why Black respondents may have viewed mask mandates less favorably. Such information could be applied to target public health measures for racialized groups in a future pandemic.

Our findings also suggest that younger people were more likely to participate in protests, which is consistent with other evidence that racial justice protests have been youth-led (Yellow Horse et al., 2021). In terms of gender, cisgender men in our sample were less likely to view mask mandates favorably, yet they were more likely to participate in protests. These are some potential explanations for why cisgender women were less likely to participate in protests than cisgender men. Notably patriotism and citizenship in the United States centers a White heterosexual masculinity as the defender and moral compass of the nation (Dickerson & Hodler, 2020). With New York State "on pause" since March 2020, many women have had to give up their careers to care for their children due to competing demands of working from home and managing their children's online learning (Schmidt, 2020). Future scholarship ought to explore the dynamics of gender, sex and sexuality in the context of pandemics, patriarchy and grassroots participation across race.

Asian respondents were less likely than white respondents to participate in protests. It is possible that members of the Asian community felt unsafe in a climate where xenophobic rhetoric was highly prevalent. Since the onset of the pandemic, many Asian Americans have suffered racial violence, including verbal harassment, vandalism and physical attacks (Center for the Study of Hate & Extremism [CSUSB], 2021). Sinophobic slurs went viral on social media when COVID-19 was referred to as "chickenpox," "kung ful," and "chinaids" with implications that China had "engineered" the virus (Li & Nicholson, 2021). In March 2020 the FBI warned of a spike in hate crimes against Asian Americans, based on the assumption that Americans would associate COVID-19 with China and Asia (Thorbecke & Zaru, 2020). Within NYC, there was a 223% increase in Asian hate crimes from the first quarter of 2020 to the first quarter of 2021 (Center for the Study of Hate & Extremism [CSUSB], 2021). Although the Asian American community in New York City is diverse in its membership, Asians constitute a smaller proportion of New York residents (15.6%) as compared to white (30.9%) and Black (20.2%) residents (New York City, 2021). Even nationally, Asian Americans tend to have lower rates of participation in protests compared to other minoritized groups (Brown et al., 2020).

Nonetheless our study indicates a high rate of protest participation by respondents, suggesting that people were willing to engage in health risks to confront state-sanctioned violence. Several public health practitioners supported racial justice protests during the stay-at-home orders, arguing that not protesting against police violence was a public health risk in itself (Kampmark, 2020). Additionally, there has been little evidence of protests resulting in spikes of COVID-19 cases (Lazer et al., 2021). It could be possible that while participants may not have supported the mandate, as was the case in our study, they may have still been wearing masks.

Our study has several limitations affecting the generalizability of the results. All data collected was self-reported with a potential for response bias and social desirability bias. Our sample was primarily composed of cisgender, heterosexual participants and hence we were unable to gauge the impact of sexual orientation on attitudes regarding the mask mandates or likelihood of participating in racial justice protests.

Our study also did not inquire about the extent to which respondents participated in the racial justice protests through social media. Online posts on Twitter, Facebook, Instagram, YouTube and other forms of social media are increasingly becoming a modality for social protest (Ince et al., 2017). Our study also did not inquire about the political viewpoints of participants, which may have also limited our findings, as attitudes regarding mask mandates are a highly politicized issue in America (Kahane, 2021). In a study of 250,000 U.S. respondents between July 2 and July 14, 2020 (just preceding the time period for our own survey data collection), mask-wearing behavior was inversely related to voting for Trump in the 2016 election (Kahane, 2021). Exploring the political motivations behind mask-wearing mandates and racial justice protest participation is an area that warrants further research, especially given the possibility for another global pandemic in the future and the reality of ongoing racial inequity in the United States.

Our findings, despite limitations, serve to identify areas for social work intervention in future pandemics with a focus on systemic racism. It is important to include social workers in pandemic preparedness planning and response, given the profession's emphasis on social and racial justice. Social workers are skilled in navigating ethical challenges as guided by the National Association of Social Workers Code of Ethics (Ross & Zerden, 2020) and the human rights framework values and principles (see, Table 3). Social workers ought to continue their efforts to strengthen human rights into the fabric of society in ways that can help inform decision-making and steward the value restructuring on which to build a racially just society (Maschi, 2016).

Consistent with CRT, the dual pandemic has brought to the forefront the voices, visibility, and rights to grassroots participation of those who are vulnerable to racism, systematic social exclusion, and other structural forms of discrimination. Of the six guiding principles to decision-making and community (see, Table 3), the principle of participation was particularly underscored in our study. As our findings suggest, some citizens felt a reaction to engage in civic participation, especially related to BLM protests. Black voices were seen and heard, loud and clear, despite the mask mandate. While the COVID-19 era has made whites aware that systemic change is needed to address racism in our society, these efforts will fail if whites reduce their commitment.

Given that social workers are overwhelmingly white women (Salsberg et al., 2017), it is important that social workers are trained (preservice and in-service) to engage in anti-oppressive social work practice. Anti-oppressive practice (AOP) acknowledges that oppression is a system of domination that denies dignity, social resources, human rights and power (Gerassi et al., 2021). AOP demands that social workers, particularly those who identify as white, are reflective and critical in challenging domination at the interpersonal and structural level (Gerassi et al., 2021). Awareness is critical to AOP and sets the stage for allyship which should not be performative in nature, but rather exhibits a willingness to be challenged, change, and act. Leveraging on CRT and AOP, clinical social workers should challenge oppressive social structures and be cognizant of the power dynamics in client/provider relationships (Sakamoto & Pitner, 2005). It is important to allow clients the space to narrate their lived experiences (Sakamoto & Pitner, 2005) to prevent the reproduction of dominant social discourse and racialized stereotyping.

Although the COVID-19 response has resulted in a surge of white allyship in confronting anti-Black racism, there is also a need for cross-racial solidarity to combat white supremacy. The racialization of COVID-19 and the concurrent surge of anti-Asian discrimination during the pandemic has raised complexities in regard to the participation of Asian Americans in racial justice protests, an issue which is further complicated by the fact that one of the police officers who stood by during Floyd's murder is of Hmong descent. This warrants a deeper investigation into whether Asian

Table 3. Human rights framework: core values and principles.

Human rights values
1) Dignity, worth, and respect of all persons
2) Intrinsic value of all persons
3) The duty to be of service to others (benevolence; do no harm).
Guiding principles to decision-making and community building
(1) Universality (inclusive of everyone)
(2) Nondiscrimination (access to rights for everyone)
(3) Indivisibility and interdependence of rights (the "right" to all rights: political, civil, social, economic, and solidarity/cultural rights)
(4) Full participation (of all citizens in their communities)
(5) Accountability (for monitoring and evaluating practices, policies, and laws)
(6) Transparency (especially of information and decision-making of leaders and governments)

American communities may be complicit in anti-Black racism or may not see the struggles of Black communities as their own. Alternatively, Asian Americans may not have engaged in protests due to the increase in anti-Asian hate crimes at that point in time. Social workers can play a role in continuing to positively engage, educate, and support Black, Asian, white, and other racial/ethnic groups in creating diverse, equitable, and inclusive social movements. Power relations amongst communities of color must also be closely examined and supported to engage in inter-racial solidarity and to dismantle white supremacy.

Conclusion

COVID-19 and racism represent epistemic threats to our society. As social workers, we identify the need to be more explicit in our actions to combat racism in various spheres of public and professional life. As educators, we are committed to engaging in anti-oppressive social work practice within social work educational systems and within social work practice settings. We recognize the need to actively question why topics regarding the reproduction of racism within the field of social work have largely gone unaddressed, or at the very least have been resisted, by the social work field since its very inception.

Disclosure statement

No potential conflict of interest was reported by the author(s).

Funding

This work was supported by the Fordham University

ORCID

Rahbel Rahman (iD) http://orcid.org/0000-0003-1065-7084
Sameena Azhar (iD) http://orcid.org/0000-0002-2249-8976
Laura J. Wernick (iD) http://orcid.org/0000-0001-6785-4658
Jordan E. DeVylder (iD) http://orcid.org/0000-0001-5873-4582
Tina Maschi (iD) http://orcid.org/0000-0002-5173-2151

References

Alsan, M., & Wanamaker, M. (2018). Tuskegee and the health of black men. *The Quarterly Journal of Economics, 133*(1), 407–455. https://doi.org/10.1093/qje/qjx029

Black Lives Matter,. (2021). *About Black Lives Matter.* Retrieved on February 27th 2021 from: https://blacklivesmatter.com/

Brown, N. E., Block, R., Jr., & Stout, C. ((Eds.). (2020). *The politics of protest: Readings on the Black Lives Matter movement.* Routledge.

Cage, M. (2020, October 16). This summer's Black Lives Matter protesters were overwhelmingly peaceful, our research finds. The Washington Post. https://www.washingtonpost.com/politics/2020/10/16/this-summers-black-lives-matter-protesters-were-overwhelming-peaceful-our-research-finds/

Center for the Study of Hate & Extremism [CSUSB]. (2021). *Report to the* Anti-Asian Prejudice & Hate Crime.

Centers for Disease Control and Prevention [CDC]. (2021a). *How to protect yourself & others.* Retrieved on March 1st 2021 from: https://www.cdc.gov/coronavirus/2019-ncov/prevent-getting-sick/prevention.html

Centers for Disease Control and Prevention [CDC]. (2021b). *Science brief: SARS-CoV-2 and potential airborne transmission.* Retrieved on March 1st 2021 from: https://www.cdc.gov/coronavirus/2019-ncov/more/scientific-brief-sars-cov-2.html

Centers for Disease Control and Prevention [CDC]. (2021c). *CDC COVID data tracker: Maps, charts, and data.* Retrieved on March 1st 2021 from: https://covid.cdc.gov/covid-data-tracker/#cases_casesper100klast7days

Centers for Disease Control and Prevention [CDC]. (2021d). *Cases, data, and surveillance*. Retrieved on 2021 March 1 from https://www.cdc.gov/coronavirus/2019-ncov/covid-data/investigations-discovery/hospitalization-death-by-race-ethnicity.html#footnote03

Çetinkaya, H. (2020). Black Lives Matter, covid-19 and the scene of politics. *The New Pretender*.

Delgado, R., & Stefancic, J. (2017). *Critical race theory: An introduction* (Vol. 20). New York University Press.

Dickerson, N., & Hodler, M. (2020). "Real men stand for our nation": Constructions of an American Nation and Anti-Kaepernick memes. *Journal of Sport and Social Issues 45* 4 , 329–357. https://doi.org/10.1177/0193723520950537

Dixson, A. D. (2018). "What's going on?": A critical race theory perspective on Black Lives Matter and activism in education. *Urban Education, 53*(2), 231–247. https://doi.org/10.1177/0042085917747115

Dreyer, B. P., Trent, M., Anderson, A. T., Askew, G. L., Boyd, R., Coker, T. R., & Mendoza, F. (2020). The death of George Floyd: Bending the arc of history toward justice for generations of children. *Pediatrics, 146*(3). https://doi.org/10.1542/peds.2020-009639

Edwards, F., Lee, H., & Esposito, M. (2019). Risk of being killed by police use of force in the United States by age, race–ethnicity, and sex. *Proceedings of the National Academy of Sciences, 116*(34), 16793–16798. https://doi.org/10.1073/pnas.1821204116

Gerassi, L. B., Klein, L. B., & Rosales, M. D. C. (2021). Moving toward critical consciousness and anti-oppressive practice approaches with people at risk of sex trafficking: Perspectives from social service providers. *Affilia*, 088610992110255. https://doi.org/10.1177/08861099211025531

Gibson, A. N., Chancellor, R. L., Cooke, N. A., Dahlen, S. P., Patin, B., & Shorish, Y. L. (2020). Struggling to breathe: COVID-19, protest and the LIS response. *Equality, Diversity and Inclusion: An International Journal, 40* (1 74–82). https://doi.org/10.1108/EDI-07-2020-0178

Gostin, L. O., Cohen, G., & Koplan, J. K. (2020). Universal masking in the United States: The role of mandates, health education, and the CDC. *JAMA, 324*(9), 837–838. https://doi.org/10.1001/jama.2020.15271

Ince, J., Rojas, F., & Davis, C. A. (2017). The social media response to Black Lives Matter: How Twitter users interact with Black Lives Matter through hashtag use. *Ethnic and Racial Studies, 40*(11), 1814–1830. https://doi.org/10.1080/01419870.2017.1334931

Jan, T. (2020, April 9). Two black men say they were kicked out of Walmart for wearing protective masks. Others worry it will happen to them. The Washington Post. https://www.washingtonpost.com/business/2020/04/09/masks-racial-profiling-walmart-coronavirus/

Jean, T. (June 16, 2020). Black lives matter: Police brutality in the era of COVID-19. *Lerner Center for Health Promotion, Syracuse University*, , https://lernercenter.syr.edu/wp-content/uploads/2020/06/Jean.pdf.

Kahane, L. H. (2021). Politicizing the Mask: Political, economic and demographic factors affecting mask wearing behavior in the USA. *Eastern Economic Journal 47* 2 , 163–183. https://doi.org/10.1057/s41302-020-00186-0

Kaiser Family Foundation [KFF]. (2021, April 28). *Latest data on COVID-19 vaccinations race/ethnicity*. https://www.kff.org/coronavirus-covid-19/issue-brief/latest-data-on-covid-19-vaccinations-race-ethnicity/

Kampmark, B. (2020). Protesting in pandemic times: COVID-19, public health, and Black Lives Matter. *Contention, 8* (2), 1–20. https://doi.org/10.3167/cont.2020.080202

Kishi, R., & Jones, S. (2020). Demonstrations & political violence in America. New data for summer 2020. https://acleddata.com/2020/09/03/demonstrations-political-violence-in-america-new-data-for-summer-2020

Krieger, N. (2020). ENOUGH: COVID-19, structural racism, police brutality, plutocracy, climate change—and time for health justice, democratic governance, and an equitable, sustainable future. *American Journal of Public Health, 110* (11), 1620–1623. https://doi.org/10.2105/AJPH.2020.305886

Lawrence, C. (2020). *Masking up: A COVID-19 face-off between Anti-mask laws and mandatory mask orders for Black Americans. Caroline Lawrence & COVID-dynamic team, masking up: A COVID-19 face-off between anti-mask laws and mandatory*. Mask Orders for Black Americans, California Law Review Online.

Lazer, D., Santillana, M., Perlis, R. H., Ognyanova, K., Baum, M., Druckman, J., & Simonson, M. (2021). *The COVID states project #10: The pandemic and the protests*. https://doi.org/10.31219/osf.io/qw43g

Li, Y., & Nicholson, H. L., Jr. (2021). When "model minorities" become "yellow peril"—Othering and the racialization of Asian Americans in the covid-19 pandemic. *Sociology Compass, 15*(2), e12849. https://doi.org/10.1111/soc4.12849

Lowe, R. H. (2020). *Policing, justice, and Black communities. Part 1: A historical overview*. The University of Texas at Austin, Institute for Urban Policy Research & Analysis.

Mahbubani, R. (2020, April 21). New Yorkers are now required to cover their faces in public, but the governor's executive order overlooks a penal code that says wearing masks counts as loitering. *Business Insider*. https://www.businessinsider.com/new-york-face-coverings-coronavirus-penal-code-wearing-masks-loitering-2020-4

Martinelli, L., Kopilaš, V., Vidmar, M., et al. (2021). Face Masks During the COVID-19 Pandemic: A Simple Protection Tool With Many Meanings. *Frontiers in Public Health, 8*, 606635. https://doi.org/10.3389/fpubh.2020.606635

Maschi, T. (2016). *Applying a human rights approach to social work research and evaluation: A rights research manifesto*. Springer Publishing.

New York City. (2021). *2020 census results for New York City: Key population & housing characteristics*. Retrieved on September 20th 2021 from: https://www1.nyc.gov/assets/planning/download/pdf/planning-level/nyc-population/census2020/dcp_2020-census-briefing-booklet-1.pdf?r=3

New York State [NYS] Governor's Office. (2020a). *Amid ongoing COVID-19 pandemic, governor Cuomo issues executive order requiring all people in New York to wear masks or face coverings in public*. Retrieved on March 6th 2021 from: https://www.governor.ny.gov/news/amid-ongoing-covid-19-pandemic-governor-cuomo-issues-executive-order-requiring-all-people-new

New York State [NYS] Governor's Office. (2020b). *Governor Cuomo Issues guidance on essential services under The 'New York State on PAUSE' Executive Order*. Retrieved on March 6th 2021 from: https://www.governor.ny.gov/news/governor-cuomo-issues-guidance-essential-services-under-new-york-state-pause-executive-order

New York State [NYS] Governor's Office. (2020c). *Governor Cuomo and Mayor de Blasio Announce Citywide Curfew in New York City will take effect beginning at 11 PM tonight*. Retrieved on March 6th 2021 from: https://www.governor.ny.gov/news/governor-cuomo-and-mayor-de-blasio-announce-citywide-curfew-new-york-city-will-take-effect

Phelan, J. C., & Link, B. G. (2015). Is racism a fundamental cause of inequalities in health? *Annual Review of Sociology, 41* (1), 311–330. https://doi.org/10.1146/annurev-soc-073014-112305

Ross, A. M., & Zerden, L. (2020). Prevention, health promotion, and social work: Aligning health and human service systems with a workforce for health. *American Journal of Public Health, 110*(S2), S186–S190. https://doi.org/10.2105/AJPH.2020.305690

Sakamoto, I., & Pitner, R. O. (2005). Use of critical consciousness in anti-oppressive social work practice: Disentangling power dynamics at personal and structural levels. *British Journal of Social Work, 35*(4), 435–452. https://doi.org/10.1093/bjsw/bch190

Salsberg, E., Quigley, L., Mehfoud, N., Acquaviva, K., Wyche, K., & Sliwa, S. (2017). *Council on social work education and national workforce initiative steering committee*. Profile of the Social Work Workforce. https://www.cswe.org/Centers-Initiatives/Initiatives/National-Workforce-Initiative/SW-Workforce-Book-FINAL-11-08-2017.aspx

Schmidt, S. (2020, May 5). *Women have been hit hardest by job losses in the pandemic. And it may only get worse. Washington Post*. https://www.washingtonpost.com/dc-md-va/2020/05/09/women-unemployment-jobless-coronavirus/

Taylor, K.-Y. (2020, May 29). Of course there are protests. The state is failing Black people. *New York Times*. https://www.nytimes.com/2020/05/29/opinion/george-floyd-minneapolis.html

Taylor, S., & Asmundson, G. J. G. (2021). Negative attitudes about face masks during the COVID-19 pandemic: The dual importance of perceived ineffectiveness and psychological reactance. *PLoS ONE, 16*(2), 1–15. https://doi.org/10.1371/journal.pone.0246317

Thorbecke, C., & Zaru, D. (2020, May 20). *Asian Americans face coronavirus "Double Whammy": Skyrocketing unemployment and discrimination. ABC News*. https://abcnews.go.com/Business/asian-americans-face-coronavirus-double-whammy-skyrocketing-unemployment/story?id=70654426

Wong, B. (2020, September 20). *Why East Asians were wearing masks long before COVID-19. HuffPost*. https://www.huffpost.com/entry/east-asian-countries-face-masks-before-covid_l_5f63a43fc5b61845586837f4

Yellow Horse, A. J., Kuo, K., Seaton, E. K., & Vargas, E. D. (2021). Asian Americans' indifference to Black Lives Matter: the role of nativity, belonging and acknowledgment of Anti-Black Racism. *Social Sciences, 10*(5), 168. https://doi.org/10.3390/socsci10050168

Model Minority Mutiny: addressing anti-Asian racism during the COVID-19 pandemic in social work

Dale Dagar Maglalang ⓘ, Smitha Rao ⓘ, Bongki Woo ⓘ and Kaipeng Wang ⓘ

ABSTRACT

The rise of anti-Asian racism during the COVID-19 pandemic signaled a lasting and ongoing history of racism in the United States. These events were a reminder to reexamine the condition of Asians and Asian Americans in the field of social work in the U.S. The purpose of this article is to support inter-solidarity movements in social work to uplift the lived experiences of Asian Americans with four recommendations: conceptualizing and positioning the Asian American identity, acknowledging the heterogeneity of the Asian American population, integrating Asian American history in the social work curriculum, and using research strategies to address anti-Asian racism uplift the experiences of Asian Americans.

Introduction

The outbreak of the COVID-19 pandemic led to an influx of anti-Asian racism in the United States (U.S.) and other parts of the world. Racist politicians such as the former President of the U.S., Donald J. Trump, racialized the virus and placed the blame on China with racist appellations such as "China Virus" and "Chinese coronavirus" (Rizzuto, 2020). As a result, racist conservatives felt emboldened to target Asians and Asian Americans[1] and incite violence. A coalition of organizations monitoring anti-Asian racism during the COVID-19 pandemic, Stop AAPI Hate, collected over 3,795 self-reported anti-Asian racism incidents from March 2020 through February 2021 (Jeung et al., 2021). Asian Americans reported experiencing violence ranging from verbal and physical harassment to vandalism of property. About 68% of Asian American women disproportionately experienced most of this reported violence (Jeung et al., 2021). Such directed violence toward Asian American women became further evident when a white supremacist murdered six Asian American women (Soon Chung Park, Hyun Jung Grant, Suncha Kim, Yong Aae Yue, Xiaojie Tan, and Daaoyou Feng) in Asian spas in Atlanta, Georgia, on March 16, 2021 (McWhirter & Bauerlein, 2021). While the COVID-19 pandemic marked an increase in anti-Asian racism, Asian Americans have experienced discrimination and racism since they immigrated to the U.S. In this paper, we provide an overview of historical events in the U.S. that preceded the current spate of racial discrimination, situate this specifically in the context of social work, and provide recommendations to social work educators, researchers, and practitioners to advance anti-racist pedagogy and practice in their work with Asian and Asian American groups.

Overview of anti-Asian racism

In the late 1800s and 1900s, the U.S. implemented a series of racialized policies. In 1885, the Page Act barred Chinese women from immigrating because of suspicions of prostitution (Kang, 2020). The Chinese Exclusion Act of 1885, the Gentleman's Agreement of 1907, and the Tydings-McDuffie Act of 1934 limited immigration and placed quotas on Chinese, Japanese, and Filipinos, respectively (Takaki, 1989). From 1942 through 1945, over 120,000 Japanese Americans were incarcerated in internment camps under the false fear of espionage during World War II (Takaki, 1989). In 1987, Navroze Mody was murdered in Jersey City, New Jersey, by the Jersey City Dot Busters (Anand, 2006). Their name derived from the *bindi* (a colored dot or point on the forehead) commonly worn by South Asian women (Anand, 2006). Anti-South Asian racism and Anti-Muslim sentiments reemerged post 9/11, where many South Asian Muslims and Sikhs were profiled and attacked by racists falsely accusing them of the 9/11 attacks (Maira, 2008). The prevalent xenophobia and racism bleeds into and affects multiple racial and ethnic minorities, as was evident when Srinivas Kuchibhotla, a South Asian immigrant, was assumed to be West Asian and asked to "get out of my country" before being shot and murdered by the gunman (Tessler et al., 2020)

Considering the ongoing public health crisis, it is essential to note that Asians have been blamed historically for the proliferation of numerous diseases and viruses. The various smallpox epidemics in San Francisco in the late 1800s and the bubonic plague in the early 1900s led public health officials to quarantine and fumigate the homes of Asian residences in Chinatown (Shah, 2001). Compared to their European counterparts entering with ease at Ellis Island, thousands of Asian immigrants entering the U.S. in San Francisco were detained for weeks, months, and sometimes years at Angel Island from 1910 through the 1940s (E. Lee, 2015). In Angel Island, Asian immigrants were subjected to physical examinations in search of "Oriental diseases," which could bar them from entering the country, and were forced to provide stool samples (E. Lee, 2015). In 1939, white doctors in Hawaii related the disease odontoclasia, i.e., tooth decay, among Filipino and Native Hawaiian children, blaming the influence of their culture and lack of assimilation to the U.S. (J.J. Kim, 2010); however, further research showed that odontoclasia resulted from poverty and the poor and limited diet afforded to Filipino and Native Hawaiian children compared to their white counterparts. The 2003 SARS outbreak, too, led to anti-Asian racism, specifically toward Chinese and Filipinos in the North American continent (Fang, 2020; Leung, 2008).

Asian Americans in social work

In the wake of the surge of anti-Asian racism in the U.S. during the pandemic, we are reminded of the enduring presence of racism experienced by Asian Americans. Within the field of social work too, Asian Americans have routinely experienced anti-Asian racism. Social workers were instrumental in the incarceration of Japanese Americans during WWII by facilitating their removal from their homes to internment camps (Park, 2015). Asian American social work students are also routinely invisibilized and often experience racial discrimination (Kwong, 2018). A qualitative study examining the experiences of Asian American social work students in the U.S. found common impressions of belittlement and second-guessing their own cultural experiences and knowledge due to being invalidated by educators or field instructors (Chung, 2008). These anti-Asian racist experiences in social work may potentially contribute to the underrepresentation of Asian Americans in the field and social work programs. According to the 2020 labor force statistics, only 3.7% of employed social workers in the country are Asian Americans, much lower than the percentage of Asian Americans (6%) among the U.S. labor force in general (Bureau of Labor Statistics U.S. Department of Labor, 2020a, 2020b). This statistic is reflected in the pipeline of Asian American social work students who comprised a minority across various social work programs in 2019: 2.2% in BSW, 3.5% in MSW, 3.3% in DSW, and 9.9% in Ph.D. programs, although those with a PhD degree is not reflective of the labor force statistics (Council on Social Work Education, 2020). Among social work faculty, 6.8% of full-time and 2.4% of part-time faculty are Asian Americans (Council on Social Work Education, 2020). It is imperative to understand further the disaggregation of these percentages by Asian American ethnic groups. While this is not a comprehensive list of all the key events in the long history of discriminatory policies and practice, this provides a glimpse of just how entrenched Anti-Asian racism has been at multiple levels including in the field of social work.

Anti-racist social work practice, education, and research for Asian Americans?

In reflection of the current racial climate in the U.S., it is critical to re-visit the experiences of Asian Americans in the U.S. and the social work profession. The rapid surge of anti-Asian racism during the COVID-19 pandemic calls for more active practice, educational, and academic responses. First, in order to address anti-Asian racism, social work scholars and practitioners need to reflect on their own biases toward people of Asian ancestry and to raise their own awareness of racial experiences of Asian Americans. Based on the understanding of racial positions of Asian American, social workers can challenge anti-Asian racism in their multiple roles. For example, social workers can develop culturally specific evidence-based interventions that can address the detrimental effects of anti-Asian discrimination. They can also promote and advocate for anti-racist policies and programs that disrupt anti-Asian racism. Social work educators can develop implicit and explicit curriculums that raise awareness of Asian American groups and their racial experiences. Social work researchers can build rigorous scientific evidence regarding the prevalence and negative impacts of anti-Asian racism and its mechanism. Below, we provide recommendations and implications for social work educators, researchers, and practitioners to uplift the lived experiences of Asian Americans and tools to address anti-Asian racism.

Conceptualize and position Asian Americans

Social work scholars must acknowledge the complexity of the Asian American identity and the positionality of Asian Americans in the U.S. Race as a social construct was conceptualized to purport a false belief of white superiority over other racial groups to justify the inequitable allocation of resources and the colonization and enslavement of people (Cooper & David, 1986). Whereas people of Asian descent have been historically described with labels, such as Oriental and Asiatics; inspired by the Black Power Movement and the fight for Ethnic Studies, Yuji Ichioka and Emma Gee founded the Asian American Political Alliance (AAPA) at the University of California, Berkeley in 1968, which first used the term Asian American (Dirlik, 2010; E. Lee, 2015). The term challenged Oriental cultural hegemony and recognized the shared history and struggles of Asian Americans (Dirlik, 2010). The term Asian American has evolved and has been used to access funding, resources, and political power. Numerous iterations have also emerged to encompass other racial groups such as Asian Pacific Islander (API), Asian American, Native Hawaiian, and Pacific Islander (AANHPI), Asian Pacific Islander, and Desi Americans (APIDA), and Asian American, Native Hawaiian, and Other Pacific Islander (AANHOPI). However, it is essential to acknowledge that Pacific Islanders and Native Hawaiians have distinct histories and experiences from Asian Americans (Kauanui, 2005). Thus, using such terms needs to be weighed carefully lest identities, groups, and lived experiences are lumped together.

Understanding the positionality of Asian Americans in the U.S. is integral in comprehending the workings of white supremacy. Extant theoretical frameworks like the Racial Triangulation Theory & Racial Position Model illustrate how Asian Americans have been placed in proximity to whiteness and racial superiority while simultaneously being othered as perpetual foreigners (C. J. Kim, 2016; Zou & Cheryan, 2017). This is operationalized through concepts such as the Model Minority Myth (MMM) that suggest that Asian Americans are academically and economically successful through hard work, therefore, implicating the inability of other nonwhite racial groups (e.g., Black, Latinx, Native Americans, etc.) to achieve similar success as Asian Americans (E. D. Wu, 2014). In fact, through the overrepresentation of some Asian groups in select fields, the MMM erases and makes invisible, many of the disparities within Asian and Asian American groups (Shih et al., 2019). The MMM also pits Asian Americans against other Black, Indigenous, and People of Color (BIPOC) communities to compete for the limited resources available to them and deters inter-solidarity organizing to challenge and eradicate white supremacy. Some Asian Americans may also internalize the MMM, which has been associated with anti-Blackness and lack of support for affirmative action (Yi & Todd, 2021), thus advancing the agenda of white supremacy, which also negatively affects Asian Americans and, in the process, disregards disparities within the Asian American population. More importantly, while we discussed the positionality of Asian Americans in the U.S., we recognize the role of Asian Americans as colonial settlers and encourage the importance of working alongside Indigenous communities to shift material wealth back to the original stewards of this land. We also acknowledge the crucial need to challenge the role of the U.S. in propagating imperialism, militarization, and (neo)colonization of countries in Asia that influence the economies, (im)migration, and well-being of Asians.

Acknowledge the heterogeneity and tensions in the Asian American population

Asian Americans are not a monolith and are composed of multitudes of Asian ethnic groups with intersecting identities. There are over 22 million Asian Americans in the U.S., the fastest-growing immigrant population in the country, with over 22 groups with origins from different parts of continental Asia (Budiman & Ruiz, 2021). Despite this growth, there is still limited data about Asian Americans. The available data either lump all Asian Americans into one group or primarily only include certain Asian ethnic groups. Many researchers and community organizations have been advocating for government sectors to oversample and disaggregate data in the Asian American population (Holland & Palaniappan, 2012). Through these efforts, we can better identify disparities within the Asian American community. For instance, in 2018, a report found that 21.7% of Asian Americans in New York City lived in poverty, higher than the white and Black population in the city (Shih et al., 2019). Several Asian ethnic groups have some of the lowest educational attainment levels in the country, with only 15% of Bhutanese and 18% of Laotians having a bachelor's degree or higher (Budiman & Ruiz, 2021). Some Asian groups like Nepalese (33%) and Burmese (46%) have lower homeownership rates compared to other racial and ethnic groups (Budiman & Ruiz, 2021). During the pandemic, 31.5% of COVID-19 deaths among nurses in the U.S. were of Filipino American descent, despite Filipino Americans only comprising 4% of nurses in the country (Akhtar, 2020). The availability of disaggregated data can help social work researchers and practitioners distinguish where research and services are needed to meet under-resourced Asian American groups' needs.

Intersecting tensions and schisms within the Asian American populations ought to be interrogated and considered to be truly anti-oppressive. For instance, much of the privilege and status among South Asian communities in the U.S. was built on white adjacency and an erasure of casteist practices within their communities. Often regarded as a success story in the U.S., a large section of South Asian Americans come from caste privilege in their home countries that contributed to their immigration as "highly skilled" workers when the first waves of immigration began in 1965 and led to a concentration of skills within these communities owing to that privilege (Chrispal et al., 2020). In fact, even before the 1965 immigration laws that marked an increase in the number of Asian immigrants into the U.S., landmark U.S. citizenship cases such as those of Bhagat Singh Thind from India and Takao Ozawa from Japan were argued based on their proximity to whiteness rather than a critique on the arbitrarily racialized immigration ceilings in the legislation (Sutherland & Supreme Court of the United States, 1922, 1923). Following closely on the heels of the Ozawa ruling that confined whiteness to the "Caucasian," Thind claimed to belong to a "high caste Hindu stock" that qualified him to be part of the Caucasian race. Therefore, the evidence of caste hierarchy seeping into a racialized hierarchy has a long history and has played a role in perpetuating white supremacy in the U.S.

Relatedly, anti-Blackness and colorism, too, are a common feature across Asian American and Asian communities and are closely connected to and white supremacy within the global immigration policy (Bashi, 2004; Dutt-Ballerstadt, 2020). Anti-Blackness runs parallel to and is entrenched in the social hierarchies within many Asian communities. Casteism and colorism, for instance, are still widely prevalent among South Asians and often carried with individuals and communities as they emigrate outside the geographies. Colorism among many ethnic and racial minorities has implications for within-group socialization, which is relevant for several sociological and public health outcomes (Burton et al., 2010). Systems of oppression often intersect in the quest for power. Therefore, tackling these inequities would empower Asian Americans to join forces to upend white supremacy and similar oppressive systems. With our responsibility to recognize the heterogeneity within Asian communities comes the imperative to interrogate oppressive systems of social hierarchies within these communities as well (Adur & Narayan, 2017).

Integrate Asian American history, experiences, and inter-solidarity movements in social work education and training

Asian Americans have a rich history and current ongoing movements that display the values of social work. Scholarship published in academic disciplines such as Ethnic Studies and Asian American Studies has shown the rich and enduring history of Asian Americans providing social welfare services and community organizing that can inform and expand social work education and research. Because of anti-Asian racism and exclusionary laws that deterred Chinese Americans from accessing social welfare services, Chinese organizations based on ancestral descent, common regions, and similar languages emerged and later coalesced to form the Consolidated Chinese Benevolent Association (CCBA; Hung, 2016). This organization provided mutual aid, financial support, employment and translational services, and legal protection from hate crimes for Chinese immigrants (Hung, 2016). Inspired by the Black Panther Party, in the late 1960s, the Red Guard Party in San Francisco was formed and adopted a 10-Point Program encompassing education, housing, healthcare, freedom from carceral and military systems, and the right to self-determination. They provided free breakfast to children in Chinatown and advocated for abolishing police in Chinatown (Maeda, 2005). In the 1960s and 1970s, Asian American students and allies advocated for fair housing and against threats of evictions for elderly and poor Asian Americans through movements such as the fight for the International Hotel (I-Hotel) in San Francisco and the International District of Seattle (Habal, 2007). Filipino American students organized to help construct the Agbayani Village in Delano, CA, a retirement home for aging Filipino American members of the United Farm Workers (UWF; Habal, 2007; Scharlin & Villanueva, 2011). Moreover, Asian American youth also organized free clinics across California's Central Valley operated by volunteer social workers and healthcare workers (Habal, 2007). Similarly, *Sakhi* ('friend' in Hindi) for South Asian Women has worked since the late 1980s on organizing, community outreach, policy advocacy, and providing culturally rooted interventions to assist survivors of domestic violence across the South Asian diaspora (Munshi et al., 2015).

Despite a few moves in the right direction from the new administration in the U.S., such as the executive order signed in the early days of the Biden presidency (The White House, 2021), concrete and institutional action toward correcting and tackling anti-Asian xenophobia and hate crime has generally been lacking. Much of the weight of reporting, documenting, building solidarity, and coalition building against anti-Asian violence is borne by Asian American and Black-led organizations. There is a long history of interracial solidarity between these groups that is often forgotten in popular discourse. Afro-Asian solidarity transpired out of an understanding of the common resistance to colonialism and imperialism fueled by white supremacy (Chang & 'Benji.,' 2020; S. J. Lee et al., 2020). Black civil rights leaders in the 1960s were vociferous in their protest against the U.S. spurred conflict in Southeast Asia. The prominent Black Liberation Movement and anti-war activist Yuri Kochiyama highlighted the need for a united voice against an oppressive system (Chang & 'Benji.,' 2020; Kochiyama et al., 2019). In recent times too, many grassroots organizations and movements such as Asians4BlackLives, South Asians for Black Lives, and Asian Prisoner Support Committee have led the way in building coalitions and solidarity among different Asian groups as well as between and in support of Asian and Black communities (Collins, 2021). Integrating and highlighting this rich organizing history of Asian Americans in providing essential services and needs of the Asian American community and building interracial and inter-group solidarity movements will help deepen social work education and training.

Use research strategies to address anti-Asian racism

Social work scholars can also play a pivotal role in addressing anti-Asian racism by accumulating more evidence that centers Asian Americans and examines the impacts of racism on them. To date, most social work literature on race and ethnicity focuses on Black and Latinx communities, yet relatively less attention was paid to Asian Americans. For example, a content analysis of articles published between 2006 and 2015 in three major social work journals found that only 22 articles attended to Asian Americans, while 51 articles focused on Black groups and 30 articles on Hispanic/Latinx Americans (Woo et al., 2018). Similarly, another study that analyzed 21 white papers of the Grand Challenges of Social Work found that only one paper mentioned Asian or Asian Americans, whereas Black and Latinx were mentioned in 12 and eight papers, respectively (Rao et al., 2021). More scholarship in social work on Asian Americans is required to shed light on the long history of discrimination against Asian Americans with tired stereotypes of being perpetual foreigners, and the rapidly growing anti-Asian violence during the COVID-19 pandemic.

Recommendations to improve social work research among Asian Americans and on anti-Asian racism include (1) oversampling of Asian Americans, (2) disaggregating Asian American subgroups, and (3) measuring various types of anti-Asian racism. First, social work researchers can expand the sample of Asian Americans in their studies. Given that large-scale state or national data can provide crucial information that guides policy, program, and funding decisions, it is important to collect data on Asian Americans in such surveys (Islam et al., 2010). However, Asian Americans have been often under-sampled in national studies (e.g., Behavioral Risk Factor Surveillance Survey) and omitted in their data reports or publications due to the small sample size (Holland & Palaniappan, 2012). Also, large-scale regional studies of Asian Americans have been centered in the West Coast, and there have been limited studies that represent Asians living in other areas in the U.S. (e.g., the Midwest, the Southwest, or the Southeast), making the current knowledge base of Asian Americans primarily driven by the data collected in certain geographic and social contexts (Islam et al., 2010). More social work researchers can oversample Asian Americans to ensure representation of this group in their study. Additional funding opportunities for data collection, the sustainability of such funding, and organizational support for studies of Asian Americans would be essential to ensure relevant data collection specific to Asian Americans.

Second, researchers need to disaggregate the data on Asian Americans by ethnic subgroups. As discussed before, similar to the social imagination, Asian Americans have often been treated as a single racial group in most studies, despite considerable heterogeneity within the racial group that may lead to different experiences of racism and vulnerability for different subgroups (Gee et al., 2009; Li, 2013). For example, according to a study that analyzed data from the National Latino and Asian American Study (NLAAS), Filipinx immigrants are more likely to report discrimination based on their race compared to other Asian subgroups, whereas Chinese and Vietnamese immigrants are more likely to report language discrimination (Li, 2013). Furthermore, this study found that, while racial and language discrimination was associated with the risk of psychiatric disorder among Filipinx immigrants, Chinese and Vietnamese were affected by vicarious racism (e.g., seeing friends mistreated due to race/ethnicity; Li, 2013). Though cumulative evidence suggests that racism negatively influences the health and well-being of Asian Americans (D. L. Lee & Ahn, 2011; C. Wu et al., 2020), aggregating them into one racial group may mask such ethnic differences. By having sufficiently powered samples of Asian subgroups, researchers can conduct within-group analyses that can provide relevant information to serve specific ethnic communities (Holland & Palaniappan, 2012; Islam et al., 2010; Srinivasan & Guillermo, 2011). The NLAAS is an important national study model that oversampled Chinese, Filipinx, and Vietnamese, but future studies can collect more diverse Asian American subgroup samples. In addition, research publications should report data for each Asian subgroup that was studied. Though small sample size may limit conducting analyses of specific subgroups, researchers can report which subgroups were included in their larger Asian American samples to avoid extrapolating the findings from a few Asian subgroups to the entire Asian American population (Holland & Palaniappan, 2012).

Third, researchers can develop and utilize measures that comprehensively capture the racism experienced by Asian Americans. Most discrimination literature of Asian Americans has utilized scales developed based on experiences of other racial/ethnic minorities. For example, the Everyday Discrimination Scale was developed based on interviews with Black communities (Williams et al., 2016). Though the scale has been widely used in Asian American literature (Gee et al., 2007; Jang et al., 2010), a study of measurement invariance tests identified the lack of invariance for one item of this scale (i.e., "people act as if they're better than you are.") that shows lower intercepts for Asians and Latinxs than for Black and whites, suggesting that researchers must be careful about using this specific item among Asians (G. Kim et al., 2014). The finding of this study highlights the importance of testing measurement equivalence of the extant discrimination scales when comparing experiences of diverse racial/ethnic groups (G. Kim et al., 2014). Social work researchers can also develop or utilize Asian-specific scales, such as the Internalization of the Model Minority Myth Measure (Yoo et al., 2010) or the Perpetual Foreigner Stereotypes Scale (Benner & Kim, 2009), to capture various aspects of anti-Asian racism in their research.

Conclusion

Even though anti-Asian racism has prevailed in the U.S. for the past two centuries, the recent rise in anti-Asian hate activities has recalibrated social workers' attention to this critical issue. Given the complex challenges inherent in anti-Asian racism, social workers should be acutely aware of Asian Americans' positionality, within-group heterogeneity, and the urgent need for more research on anti-Asian racism and resultant social and health disparities. After reviewing the historical and empirical literature on anti-Asian racism, and the history of organizing and solidarity movements led by Asian groups, we offer several recommendations for future social work education and research in this area. First, it is critical for social work researchers and educators to re-visit the detrimental consequences of Anti-Asian racism and its manifestations, such as the MMM, on both Asian Americans and other racially/ethnically minoritized populations. The MMM has severely obscured the disparities of health, resources, and opportunities within the Asian American population. It has also neglected the heterogeneous racism and disparities experienced by different Asian ethnic groups and deprecated other racial minorities, for living in poverty by disregarding the existence of white supremacy and systemic racism. As the COVID-19 pandemic has revealed the racism and oppression that Asian Americans continue to face, social workers should be conscious of how MMM exacerbates disparities in this population and develop cultural and community responsive interventions to reduce and eradicate anti-Asian racism. Second, despite the important role of Asian Americans in the history of social work in the U.S., Asian American social workers remain underrepresented in the foundation of social work education and its workforce. Research on Asian Americans, especially discussion of disparities, remains limited at the BSW, MSW, and doctoral levels. This limited attention aggravates the MMM, contributing to Asian Americans' low social work representation compared to other helping professions. Finally, social work education should incorporate Asian American history and ongoing interracial solidarity with other oppressed and exploited communities in its pedagogy. Social work education, research, and practice need to further engage Asian American scholars and practitioners to prepare social work professionals to commit to anti-racism, racial justice, and liberation for all.

Note

1. We use the term Asian Americans to broadly denote Asians in the U.S., that is those who are American citizens, as well as members of the Asian diaspora who are here on a visa or are undocumented. The term Model Minority Mutiny was coined by Soya Jung (2014) to be in community with the liberation of BIPOC communities and challenge white supremacy.

Disclosure statement

No potential conflict of interest was reported by the author(s).

ORCID

Dale Dagar Maglalang (iD) http://orcid.org/0000-0001-8909-0193
Smitha Rao (iD) http://orcid.org/0000-0002-6788-001X
Bongki Woo (iD) http://orcid.org/0000-0002-2528-8966
Kaipeng Wang (iD) http://orcid.org/0000-0002-2206-4318

References

Adur, S. M., & Narayan, A. (2017). Stories of dalit diaspora: Migration, life narratives, and caste in the US. *Biography - An Interdisciplinary Quarterly, 40*(1), 244–264. https://doi.org/10.1353/bio.2017.0011

Akhtar, A. (2020). *Filipinos make up 4% of nurses in the US, but 31.5% of nurse deaths from COVID-19.* Business Insider. https://www.businessinsider.com/filipinos-make-up-disproportionate-covid-19-nurse-deaths-2020-9

Anand, V. Z. J. (2006). The dotbuster effect on Indo-American immigrants. *Journal of Immigrant & Refugee Studies, 4* (1), 111–113. https://doi.org/10.1300/J500v04n01_08

Bashi, V. (2004). Globalized anti-blackness: Transnationalizing Western immigration law, policy and practice. *Ethnic and Racial Studies, 27*(4), 584–606. https://doi.org/10.1080/01491987042000216726

Benner, A. D., & Kim, S. Y. (2009). Intergenerational experiences of discrimination in Chinese American families: Influences of socialization and stress. *Journal of Marriage and Family, 71*(4), 862–877. https://doi.org/10.1111/J.1741-3737.2009.00640.X

Budiman, A., & Ruiz, N. G. (2021). *Key facts about Asian origin groups in the U.S.* Pew Research Center. https://www.pewresearch.org/fact-tank/2021/04/29/key-facts-about-asian-origin-groups-in-the-u-s/

Bureau of Labor Statistics U.S. Department of Labor. (2020a). *2020 Annual Averages - Employed persons by detailed occupation, sex, race, and Hispanic or Latino ethnicity.* https://www.bls.gov/cps/cpsaat11.htm

Bureau of Labor Statistics U.S. Department of Labor. (2020b). *Labor force characteristics by race and ethnicity, 2019.* https://www.bls.gov/opub/reports/race-and-ethnicity/2019/home.htm

Burton, L. M., Bonilla-Silva, E., Ray, V., Buckelew, R., & Freeman, E. H. (2010). Critical race theories, colorism, and the decade's research on families of color. *Journal of Marriage and Family, 72*(3), 440–459. https://doi.org/10.1111/J.1741-3737.2010.00712.X

Chang, B., & 'Benji.' (2020). From 'illmatic' to 'Kung Flu': Black and Asian solidarity, activism, and pedagogies in the Covid-19 era. *Postdigital Science and Education, 2*(3), 741–756. https://doi.org/10.1007/s42438-020-00183-8

Chrispal, S., Bapuji, H., & Zietsma, C. (2020). Caste and organization studies: Our silence makes us complicit. *Organizational Studies, 42*(9), 1501–1515. https://doi.org/10.1177/0170840620964038

Chung, I. (2008). Bridging professional and cultural values and norms. *Journal of Teaching in Social Work, 26*(1–2), 93–110. https://doi.org/10.1300/J067V26N01_06

Collins, B. (2021). *Oakland organizations working toward solidarity between Black and Asian communities.* Voices of the People of Oakland. https://oaklandvoices.us/2021/04/02/oakland-organizations-working-toward-solidarity-between-black-and-asian-communities/

Cooper, R., & David, R. (1986). The biological concept of race and its application to public health and epidemiology. *Journal of Health Politics, Policy and Law, 11*(1), 97–116. https://doi.org/10.1215/03616878-11-1-97

Council on Social Work Education. (2020). *2019 statistics on social work education in the United States.* https://www.cswe.org/getattachment/Research-Statistics/2019-Annual-Statistics-on-Social-Work-Education-in-the-United-States-Final-(1).pdf.aspx

Dirlik, A. (2010). Asians on the rim: Transnational capital and local community in the making of contemporary Asian America from Asian American. In J. Y. S. Wu & T. C. Chen (Eds.), *Asian American studies now: A critical reader* (pp. 515–539). Rutgers University Press.

Dutt-Ballerstadt, R. (2020, June 26). *Colonized loyalty: Asian American anti-Blackness and complicity.* Truthout. https://truthout.org/articles/colonized-loyalty-asian-american-anti-blackness-and-complicity/

Fang, J. (2020).*The Washington Post*. The Washington Post. https://www.washingtonpost.com/outlook/2020/02/04/2003-sars-outbreak-fueled-anti-asian-racism-this-pandemic-doesnt-have/

Gee, G. C., Ro, A., Shariff-Marco, S., & Chae, D. (2009). Racial discrimination and health among Asian Americans: Evidence, assessment, and directions for future research. *Epidemiologic Reviews*, 31(1), 130–151. https://doi.org/10.1093/epirev/mxp009

Gee, G. C., Spencer, M., Chen, J., Yip, T., & Takeuchi, D. T. (2007). The association between self-reported racial discrimination and 12-month DSM-IV mental disorders among Asian Americans nationwide. *Social Science & Medicine*, 64(10), 1984–1996. https://doi.org/10.1016/j.socscimed.2007.02.013

Habal, E. (2007). *San Francisco's international hotel: Mobilizing the Filipino American community in the anti-eviction movement*. Temple University Press.

Holland, A. T., & Palaniappan, L. P. (2012). Problems with the collection and interpretation of Asian-American health data: Omission, aggregation, and extrapolation. *Annals of Epidemiology*, 22(6), 397–405. https://doi.org/10.1016/j.annepidem.2012.04.001

Hung, Y.-J. (2016). Transnational and local-focus ethnic networks: The development of two types of Chinese social organizations in the San Gabriel Valley. *Southern California Quarterly*, 98(2), 194–229. https://doi.org/10.1525/ucpsocal.2016.98.2.194

Islam, N. S., Khan, S., Kwon, S., Jang, D., Ro, M., & Trinh-Shevrin, C. (2010). Methodological issues in the collection, analysis, and reporting of granular data in Asian American populations: Historical challenges and potential solutions. *Journal of Health Care for the Poor and Underserved*, 21(4), 1381. https://doi.org/10.1353/HPU.2010.0939

Jang, Y., Chiriboga, D. A., Kim, G., & Rhew, S. (2010). Perceived discrimination, sense of control, and depressive symptoms among Korean American older adults. *Asian American Journal of Psychology*, 1(2), 129–135. https://doi.org/10.1037/A0019967

Jeung, R., Horse, A. Y., Popovic, T., & Lim, R. (2021). *Stop AAPI Hate national report*: 3/19/20- 2/28/21. Stop AAPI Hate. https://stopaapihate.org/2020-2021-national-report/

Jung, S. 2014. The racial justice movement needs a Model Minority mutiny. *Race Files*. https://www.racefiles.com/2014/10/13/model-minority-mutiny/

Kang, M. (2020). Reproducing Asian American studies: Rethinking Asian exclusion as reproductive exclusion. *Amerasia Journal*, 46(2), 136–146. https://doi.org/10.1080/00447471.2020.1840319

Kauanui, J. K. (2005). Asian American studies and the "Pacific question. In K. A. Ono (Ed.), *Asian American studies after critical mass* (pp. 123–143). Blackwell Publishing Ltd.

Kim, J. J. (2010). Experimental encounters: Filipino and Hawaiian bodies in the U.S. imperial invention of Odontoclasia, 1928-1946. *American Quarterly*, 62(3), 523–546. https://doi.org/10.1353/aq.2010.0001

Kim, C. J. (2016). The racial triangulation of Asian Americans. *Politics & Society*, 27(1), 105–138. https://doi.org/10.1177/0032329299027001005

Kim, G., Sellbom, M., & Ford, K. L. (2014). Race/ethnicity and measurement equivalence of the everyday discrimination scale. *Psychological Assessment*, 26(3), 892–900. https://doi.org/10.1037/A0036431

Kochiyama, Y., Huggins, E., & Kao, M. U. (2019). "Stirrin' waters' 'n buildin' bridges: A conversation with Ericka Huggins and Yuri Kochiyama. *Amerasia Journal*, 35(1), 140–167. https://doi.org/10.17953/AMER.35.1.004J1162N8646161

Kwong, K. (2018). Career choice, barriers, and prospects of Asian American social workers. *International Journal of Higher Education*, 7(6), 1–12. https://doi.org/10.5430/ijhe.v7n6p1

Lee, E. (2015). *The making of Asian America: A history*. Simon & Schuster Paperbacks.

Lee, D. L., & Ahn, S. (2011). Racial discrimination and Asian mental health: A meta-analysis. *The Counseling Psychologist*, 39(3), 463–489. https://doi.org/10.1177/0011000010381791

Lee, S. J., Xiong, C. P., Pheng, L. M., & Vang, M. N. (2020). "Asians for Black lives, not Asians for Asians": Building Southeast Asian American and Black solidarity. *Anthropology & Education Quarterly*, 51(4), 405–421. https://doi.org/10.1111/aeq.12354

Leung, C. (2008). The yellow peril revisited: The impact of SARS on Chinese and Southeast Asian communities. *Resources for Feminist Research*, 33(1–2), 135–149. https://link.gale.com/apps/doc/A195680111/LitRC?u=prov98893&sid=bookmark-LitRC&xid=3d223cc8

Li, M. (2013). Discrimination and psychiatric disorder among Asian American immigrants: A national analysis by subgroups. *Journal of Immigrant and Minority Health*, 16(6), 1157–1166. https://doi.org/10.1007/S10903-013-9920-7

Maeda, D. J. (2005). Black panthers, Red guards, and Chinamen: Constructing Asian American identity through performing blackness. *American Quarterly*, 57(4), 1079–1103. https://doi.org/10.1353/aq.2006.0012

Maira, S. (2008). Flexible citizenship/flexible empire: South Asian Muslim youth in post-9/11 America. *American Quarterly*, 60(3), 697–720. https://doi.org/10.1353/aq.0.0030

McWhirter, C., & Bauerlein, V. (2021, March 19). Atlanta spa shootings: Officials release victims' names. *The Wall Street Journal*. https://www.wsj.com/articles/atlanta-spa-shootings-officials-release-victims-names-11616176366

Munshi, S., Nancherla, B., & Jayasinghe, T. (2015). Building towards transformative justice at Sakhi for South Asian women. *Race & Social Justice Law Review*, 5(2), 421–435. http://repository.law.miami.edu/umrsjlr/vol5/iss2/18

Park, Y. (2015). Facilitating injustice: Tracing the role of social workers in the World War II internment of Japanese Americans. *Social Services Review, 82*(3), 447–483. https://doi.org/10.1086/592361

Rao, S., Woo, B., Maglalang, D. D., Bartholomew, M., Cano, M., Harris, A., & Tucker, T. B. (2021). Race and ethnicity in the social work grand challenges. *Social Work, 66*(1), 9–17. https://doi.org/10.1093/sw/swaa053

Rizzuto, M. (2020, March 17). *U.S. politicians exploit coronavirus fears with anti-Chinese dog whistles.* Atlantic Council. https://medium.com/dfrlab/u-s-politicians-exploit-coronavirus-fears-with-anti-chinese-dog-whistles-ff61c9d7e458

Scharlin, C., & Villanueva, L. (2011). *Philip Vera Cruz: A personal history of Filipino immigrants and the farmworkers movement.* University of Washington Press.

Shah, N. (2001). *Contagious divides: Epidemics and race in San Francisco's Chinatown.* University of California Press.

Shih, K. Y., Chang, T.-F., & Chen, S.-Y. (2019). Impacts of the Model Minority Myth on Asian American individuals and families: Social justice and critical race feminist perspectives. *Journal of Family Theory & Review, 11*(3), 412–428. https://doi.org/10.1111/JFTR.12342

Srinivasan, S., & Guillermo, T. (2011). Toward improved health: Disaggregating Asian American and Native Hawaiian/Pacific Islander data. *American Journal of Public Health, 90*(11), 1731–1734. https://doi.org/10.2105/AJPH.90.11.1731

Sutherland, G., Supreme Court of the United States. (1922). *U.S. reports: Ozawa v. United States, 260 U.S. 178.* In Library of Congress Washington, D.C. 20540 USA. https://www.loc.gov/item/usrep260178/

Sutherland, G., Supreme Court of the United States. (1923). *U.S. reports: United States v. Thind, 261 U.S. 204.* In Library of Congress Washington, D.C. 20540 USA. https://www.loc.gov/item/usrep261204/

Takaki, R. (1989). *Strangers from a different shore: A history of Asian Americans.* Penguin Books.

Tessler, H., Choi, M., & Kao, G. (2020). The anxiety of being Asian American: Hate crimes and negative biases during the COVID-19 pandemic. *American Journal of Criminal Justice, 45*(4), 636–646. https://doi.org/10.1007/s12103-020-09541-5

The White House. (2021, January 26). *Memorandum condemning and combating racism, xenophobia, and intolerance against Asian Americans and Pacific Islanders in the United States.* https://www.whitehouse.gov/briefing-room/presidential-actions/2021/01/26/memorandum-condemning-and-combating-racism-xenophobia-and-intolerance-against-asian-americans-and-pacific-islanders-in-the-united-states/

Williams, D. R., Yu, Y., Jackson, J. S., & Anderson, N. B. (2016). Racial differences in physical and mental health: Socio-economic status, stress and discrimination. *Journal of Health Psychology, 2*(3), 335–351. https://doi.org/10.1177/135910539700200305

Woo, B., Figuereo, V., Rosales, R., Wang, K., & Sabur, K. (2018). Where Is race and ethnicity in social work? A content analysis. *Social Work Research, 42*(3), 180–186. https://doi.org/10.1093/swr/svy010

Wu, E. D. (2014). *The color of success.* Princeton University Press.

Wu, C., Qian, Y., & Wilkes, R. (2020). Anti-Asian discrimination and the Asian-white mental health gap during COVID-19. *Ethnic and Racial Studies, 44*(5), 819–835. https://doi.org/10.1080/01419870.2020.1851739

Yi, J., & Todd, N. R. (2021). Internalized Model Minority Myth among Asian Americans: Links to anti-Black attitudes and opposition to affirmative action. *Cultural Diversity and Ethnic Minority Psychology, 27*(4), 569–578. https://doi.org/10.1037/CDP0000448

Yoo, H. C., Burrola, K. S., & Steger, M. F. (2010). A preliminary report on a new measure: Internalization of the Model Minority Myth Measure (IM-4) and its psychological correlates among Asian American college students. *Journal of Counseling Psychology, 57*(1), 114–127. https://doi.org/10.1037/A0017871

Zou, L. X., & Cheryan, S. (2017). Two axes of subordination: A new model of racial position. *Journal of Personality and Social Psychology, 112*(5), 696–717. https://doi.org/10.1037/PSPA0000080

Conceptualizing anti-Asian racism in Canada during the COVID-19 pandemic: a call for action to social workers

Kedi Zhao 🆔, Carolyn O'Connor, Trish Lenz and Lin Fang

ABSTRACT

Anti-Asian racism in Canada has emerged from the COVID-19 pandemic and become more rampant. This article integrates Canadian postcolonialism, a critique of Canadian multiculturalism, and a framework of intergroup prejudice to conceptualize the covert anti-Asian racism that is entrenched in Canadian society. How COVID-19 exposes and "legitimizes" anti-Asian racism is further analyzed and included in this conceptualization. This conceptualization also includes social workers' leading roles in combating anti-Asian racism through reforming and integrating client interventions, cultural policy, social context, and offers directions that can guide future social work research and practice in improving social justice during this crisis.

Introduction

COVID-19 was first detected in China at the end of 2019 and was declared as a global pandemic by the World Health Organization on the 11th of March 2020 (World Health Organization, n.d.). Canada and the United States have been severely impacted by the COVID-19 pandemic due to the plummeting economy and the precarious public health conditions (Haley, 2020). Underrepresented cultural groups (i.e., Black, Indigenous populations, and people of color) are in a particularly adverse situation (e.g., high unemployment rates, limited access to healthcare) amid the pandemic (Canadian Union of Public Employees, 2020; Gordon, 2020). Along with the COVID-19 pandemic, awareness of anti-Black racism, ignited by the death of Mr. George Floyd, became another "pandemic" (Laurencin & Walker, 2020) alongside ongoing protests demonstrating people's determination in combating systemic racism (Wallis, 2020). Amid dual pandemics, anti-Asian racism has become overt in North America and has escalated with the tragic shooting that targeted Asian women in Atlanta (Jackson, 2021). In Canada, the Chinese Canadian National Council Toronto Chapter (CCNCTO) collected 1,150 anti-Asian racism events by February 2021, noting that the per capita rate of anti-Asian racism in Canada is higher than that of the US (Karamali, 2021). The methods of anti-Asian racism in Canada are varied and include verbal and physical attacks, inappropriate jokes, and racist graffiti (Flanagan, 2020). These racist incidents have negatively impacted the well-being of Asian individuals in Canada, as increased rates of physical health issues (e.g., physical harm or injury; Zhou, 2020) and mental health issues such as fear, anxiety, and depression (Gao, 2020) have been reported. Asian communities must therefore not only contend with the risk of exposure to the virus but must also cope with the ancillary risk and burden of victimization (Tessler et al., 2020). Asian communities in Canada have been making efforts (e.g., organizing rallies against anti-Asian racism) to improve this disadvantageous situation and promote social justice (Tsekouras, 2021).

It is noted that anti-Asian racism is not new; it is rooted in the Canadian white colonial past, as the first generation of Chinese and Japanese immigrants were regarded as inferior and treated unequally with a head tax and internment (Law, 2018; J. J. M. Lee, 1999). In the contemporary multicultural context, anti-Asian racism has reemerged (e.g., the spread of the SARS virus in 2003) and perpetuated the "yellow peril" myth (Keil & Ali, 2006; Li & Nicholson, 2021). In response to anti-Asian racism emerging from the COVID-19 pandemic, several empirical studies have been conducted to explore the effects of anti-Asian racism on different aspects of Asian individuals' general health and well-being (e.g., mental health; Lee & Waters, 2020; Wu et al., 2020). However, timely conceptual pieces that delve into the Canadian context and reflect the unique influences of the pandemic are also urgently needed to help us understand anti-Asian racism in a comprehensive manner. Hence, this article conceptualizes anti-Asian racism by integrating Canadian postcolonial and multicultural theories within a framework of intergroup prejudice. The role of COVID-19, as an infectious disease, in exacerbating Anti-Asian racism is also included. Based on the above conceptualization, how social workers can better serve Asian individuals amid anti-Asian racism caused by COVID-19, and how can they further policy improvements and reform the postcolonial social context, are also explored. This conceptualization aims to advance future social work research and practice and contribute to tangible representations of social cohesion and social justice in Canada.

Canadian postcolonialism

Emerging from the end of the Second World War (Arı & Ak, 2019; Christophers, 2007; Chung, 2012), postcolonialism is a term used to describe the contemporary global context of former colonizers and colonies amid the anti-colonialism and independent movements (Hiddleston, 2014). Postcolonialism does not have a universal definition but can be understood as a set of discussions across different disciplines regarding the political, economic, and cultural relations during the decolonialization process (Arı & Ak, 2019; Burney, 2012; Hiddleston, 2014). Internationally, postcolonialism can be referred to as a process of previous colonies worldwide seeking independence from former colonial powers (e.g., abandoning political systems established by former colonial powers; Loomba, 1998). However, postcolonialism does not mean that colonialism has come to an end or that its influences have disappeared (Ahmed, 2000; Nair, 2017); rather, it indicates that colonialism continues to have an implicit impact on global cultural orders (Bhatia & Ram, 2001). Domestically, postcolonialism also represents the sustained resistance by underrepresented cultural groups (e.g., Indigenous groups) on the powerful lasting effects of colonialism in former colonized states, which forms a dynamic hybrid between colonization and decolonialization (R. J. C. Young, 2016). In the social work discipline, there has also been a surge of decolonization discussion in recent decades. For example, Rowe et al. (2015) reflected on Indigenous approaches to better guide future social work research; Sakamoto (2007) explored how the anti-oppressive approach can be applied to social work practice in immigration and settlement services. However, despite these decolonization movements and discussions that are making progress in shattering previous colonial order, enormous influences from white supremacy and colonialism remain, particularly in culture (Wang, 2018). Cultural influences are the legacy of colonialism and are demonstrated through cultural norms, values, and policies (Paquette et al., 2017). How these postcolonial cultural influences have been enacted in Canada and affect its cultural structure in society is discussed further below.

Canada is in a unique and complicated position regarding postcolonialism. Unlike other former colonies that obtained their independence through continuous resistance or even violent revolutions (Ahlman, 2010), Canada took the initiative to sever ties with notorious white superiority and supremacy from Nazi Germany and to reform their societal structure to accommodate other cultural groups (Thobani, 2007). By doing this, the white dominance (i.e., white British and French settlers) embedded in previous colonial systems was preserved (Pillay, 2015) and reframed to accommodate the decolonization discourse that is taking place around the world (Conway, 2018). This reform perpetuates Canada's international image in the postcolonial era as an open-minded and inclusive country

(Kymlicka, 2004; Matthews, 2006), paving the way for Canada to connect with newly independent countries (Thobani, 2007). However, previously overt racist discourses (e.g., racial inferiority or superiority) did not completely disappear; instead, they have transformed into a covert cultural discussion featuring "advanced" white English and French culture (Thobani, 2007). In other words, the decolonization of Canadian society within this postcolonial context did not fundamentally alter the established colonial structure (Lawrence & Dua, 2005). The processes of decolonization and anti-racism continue to be designed and implemented from the perspective of white settlers (Simpson et al., 2011). For instance, although Indigenous groups and immigrants from other cultures are obtaining increased individual and civil rights (e.g., political participation) to participate in postcolonial Canadian society (Kymlicka, 1995), this participation is still within the unequal settlers' society and complies with the rules established by white settlers (Greensmith, 2015). This reality reflects the ongoing colonization in the postcolonial context (Lawrence & Dua, 2005). Kumar (2009) depicts how many Indigenous communities have participated in Canadian society through educational and political systems created by white settlers; as a result, some have begun to adopt a perspective of white settlers regarding Canadian history and society. This is what Kumar (2009) referred to as "humanism" (p. 45) rooted in the settlers' society and used by European settlers to assimilate Indigenous populations and eliminate their culture. This assimilation process is not unique to Indigenous populations. In immigrant populations, cultural colonization is (re)enacted in the mini-malization of Indigenous history and culture within the Canadian immigration system (Dmytriw, 2016). It is also evident in the emphasis on familiarity with English- or French-Canadian culture (e.g., speaking English and French; E. Lee, 2013) as well as obtaining the "Canadian experience" (i.e., Canadian education and work experience; Bhuyan et al., 2017; Buzdugan & Halli, 2009) in the immigration selection process. Therefore, from the outset, immigrants are complicit with the white settler narrative as it is embedded within the immigration process, further solidifying the current form of decolonization controlled by white settlers and perpetuating the marginalization of Indigenous populations (Lawrence & Dua, 2005; E. Lee, 2013; MacDonald, 2014). This postcolonial discussion of the remaining cultural colonial power in Canadian society paves the way for further analyses regarding specific cultural policies in Canada.

Postcolonial critique of Canadian multiculturalism

Consistent with the discussion of the macro postcolonial context, cultural policies in Canada reflect these continuous conflicts between decolonialization and enduring colonization (Paquette et al., 2017). Prior to the multiculturalism policy in 1988, Canadian society had overtly operated as a "white" country in which immigrants from European countries with Caucasian descents were prioritized (Matthews, 2006). Asian immigrants (e.g., Chinese, Japanese), on the other hand, were excluded and not treated equally (e.g., the head tax imposed on Chinese immigrants; J. M. Lee, 1999). As the anti-colonialism and independent movements gradually became the force that disintegrated the colonial order and liberated people living in colonies (Pierce, 2009), the multiculturalism policy was officially adopted in Canada to denounce the previously established colonial order (Paquette et al., 2017; Pillay, 2015). A series of minority rights (e.g., polyethnic rights) were given to immigrants and members of the Indigenous population to ensure their social participation and ameliorate their integration in Canada (Kymlicka, 1995). However, remnants of colonialism can still be traced and identified within this cultural policy (Day, 2000).

Thobani (2007) critically evaluated Canadian multiculturalism and stated that this policy does not fundamentally alter the colonial order established by white supremacy. Instead, the "superior" status of white Canadian culture was further strengthened (Tuck & Gaztambide-Fernandez, 2013). Specifically, multiculturalism acknowledges the dominance of white English and French culture in the Canadian social context, legitimizes its status in Canadian history, and reframes it as an open and tolerant culture that welcomes other cultures via immigrants from third world countries (MacDonald, 2014; Nagra & Peng, 2013; Pillay, 2015). Multiculturalism ensures equality for these immigrants to keep their identity and practice their culture (Law, 2018). However, this equality is based on the premise that every immigrant is required to acquire and adopt dominant white English or French cultures in Canada in order to be regarded as an integrated and successful member of Canadian society (Day, 2000; Wayland, 1997). Even though their own cultures are no longer overtly represented as inferior, the implicit inequality still exists, as the white English and French cultures are used as the norm by which to evaluate immigrants in Canada (Haque, 2012; E. Lee, 2013). Through reframing immigrants' failure to integrate into Canadian society as a personal failure to acquire aspects of Canadian culture (e.g., learning the English or French languages; Thobani, 2007), multiculturalism suppresses immigrants from other cultures and minimizes the possibility of immigrants challenging or significantly altering the current cultural hierarchy controlled by white English and French cultures (St. Denis, 2011; Thobani, 2007). Immigrants who adapt well to the dominant culture become stable human and social capital to Canadian society (Kymlicka, 2013; McLaren & Dyck, 2004), as they can steadily supply the Canadian labor market in the long term (Kustec, 2012). The integration of these immigrants also strengthens the current system of unequal multiculturalism in Canada, as they contribute to cultural diversity while also adhering to the conventions of English- and French-Canadian culture (Vernon, 2016). This interpretation is consistent with the above postcolonialism discussion, in which colonial power still exerts enormous influence on reshaping relations among different cultural groups (Wang, 2018). Multiculturalism policy facilitates the colonial remnants in Canadian society by promoting a notion of cultural diversity that is managed by implicit white supremacy (Tuck & Gaztambide-Fernandez, 2013).

Therefore, while Asians have been integrated into Canada as human and social capital, in addition to their promotion of multiculturalism, the cultural inequality grounded in multiculturalism and postcolonialism has put them in a vulnerable position. The current COVID-19 pandemic and consequent discrimination targeting Asians highlight this position within an unequal cultural hierarchy.

A framework to explain intergroup prejudice

Intergroup prejudice refers to the degrading and disparaging perceptions by some social groups toward other social groups in society (Dovidio & Gaertner, 1999; Pettigrew, 2008). It can be expressed in different forms, such as implicit bias and explicit hatred (Molina et al., 2016; Pettigrew, 2008). Abrams (2010) reviewed different theories and empirical studies of intergroup prejudice and presented a synthesized framework for understanding this phenomenon. According to Abrams (2010), this framework consists of four components: the first is the intergroup context in society, which has direct influences on the other three components: the bases, manifestations, and engagement of prejudice. Bases of prejudice affect manifestations, and manifestations affect engagement. The results from the engagement of prejudice can inform the bases of prejudice and form an iterative circle within the social context.

The intergroup context explains the importance of the social context in shaping how individuals or social groups form prejudices toward others (Molina et al., 2016). For instance, threats (Stephan & Stephan, 2000), group sizes (Oliver & Wong, 2003), and power (Schmid & Amodio, 2017) are factors mentioned by Abrams (2010) in the intergroup context. Specifically, Asian individuals are being mistakenly interpreted as the physical embodiment of the virus and therefore, are perceived to pose health threats to other social groups (Dhanani & Franz, 2020; Tessler et al., 2020). In addition, as Asian Canadians increase in numbers (Cao & Dehoorne, 2011), their enlarged presence fuels this ungrounded threat. Furthermore, as minorities, Asian groups experience unequal access to social power when compared to dominant white Canadians (Little, 2016), which can further intergroup prejudice. The bases of prejudice reveal how prejudice is formed from a psychological stance. Specifically, Abrams (2010) determined that different social values, social categorization, and stereotyping are the three main bases of prejudice. These bases of prejudice can also be used to explain the discrimination against Asian individuals during the pandemic. For example, at the beginning of the pandemic, Asian individuals often chose to wear face masks as a precaution, while face masks were not accepted by other groups at that time (I. Young, 2020); as a result, value differences regarding the use of face masks became a potential source of prejudice toward Asians.

Social categorization also fuels discrimination toward Asians. Unsubstantiated or false statements, as well as verbal and physical violence that associated this novel virus with individuals with Chinese descent (Flanagan, 2020; Jack-Davies, 2020; Zhou, 2020), indicated social categorization based on race and culture (Duckitt, 2003) and furthered the degree of discrimination targeting Asian individuals. With respect to stereotyping, one example forwarded by Abrams (2010) depicts that Westerners often cannot distinguish between different Asian groups whose cultural backgrounds are entirely different; as a result, this issue affects all Asian-presenting Canadians. This problematic stereotyping homogenizes Asian groups and assigns blame for their "savage" eating habits (e.g., eating bats) and "primitive" lifestyles (Gee et al., 2020; Li & Nicholson, 2021). These inaccurate stereotypes, juxtaposed against the characterization of white bodies as pure and innocent (Tessler et al., 2020), further degrade the image of Asians in Canadian society. The manifestations demonstrate that prejudice can culminate in different forms, including infra-humanization and language as presented by Abrams (2010). Infra-humanization is a subtle process whereby individuals typically view their own groups as more important and devalue out-groups (Haslam & Loughnan, 2014). Cultural groups with high social status and power often use implicit discriminatory comments to refer to cultural groups that have less social power in negative and simple ways (Croom, 2015). Utilizing these analyses, during the pandemic, Asian communities might experience infra-humanization imposed by other cultural groups (Roberto et al., 2020) and are described by different types of prejudice (e.g., racist slurs) by cultural groups with more social powers (Dhanani & Franz, 2020). The fourth part of Abrams (2010) framework is engagement of prejudice. Specifically, this segment focuses on intergroup contact; for instance, intergroup prejudice can be reduced by minimizing psychological distance between groups through intergroup contact and thus, fostering positive attitudes to out-groups (Molina et al., 2016). These discussions about intergroup contact indicate that discrimination toward Asians might be advanced by the lack of contact between Asians and other social groups, leading to further segregation (Enos & Celaya, 2018).

Overall, this framework provides a detailed description of how intergroup prejudice manifests. These accounts are consistent with the Canadian postcolonial and multicultural discussions, where the imbalance of cultural powers impacts intergroup interactions and may generate further discrimination, especially during the COVID-19 pandemic.

The role of COVID-19 in triggering anti-Asian racism

As mentioned in the analyses above, the unequal cultural hierarchy shaped by the postcolonial context and Canadian cultural policy makes anti-Asian racism readily apparent. Societal reactions to COVID-19 can be understood as an accelerant in exposing the covert unequal cultural hierarchy in Canadian multiculturalism, enabling anti-Asian racism, and "legitimizing" this hierarchy. During the pandemic, COVID-19 was viewed by other social groups as a foreign disease from Asian countries and a symbol attached to the Asian culture, further segregating Asian individuals from Canadian society by perpetuating their "otherness" and "foreignness" (Jack-Davies, 2020; Kibria, 2000; Roberto et al., 2020; Tessler et al., 2020). The ancient "yellow peril" discourse thus resurfaces and escalates (Li & Nicholson, 2021), illustrated through both covert cultural blame and overt racial blame. Asian cultures are interpreted as "inferior" by other cultural groups (Gee et al., 2020) while Asian individuals are regarded as "virus carriers" who are blamed for causing the pandemic and subsequently, making other groups suffer from this crisis (Das, 2020; Li & Nicholson, 2021). Moreover, differing from previous infectious diseases, the current COVID-19 pandemic contains more political meanings that further incites anti-Asian racism in North America, generating more detrimental consequences. Specifically, increasing economic and political tensions between China and Western countries show how the postcolonial order led by the West, especially the USA, has been gradually challenged by a rising

China. Derogatory comments made by politicians (e.g., Donald Trump referred COVID-19 as the "Chinese virus"; Viala-Gaudefroy & Lindaman, 2020), show how COVID-19 has been politicized and used to obtain political gains. This situation reflects power dynamics in the contemporary postcolonial context and fuels anti-Asian racism in Western countries.

Discussion

Based on the theoretical analyses from the macro postcolonial social context, meso multicultural policy, and micro intergroup contact, we put forward a conceptual model (Figure 1) to understand Anti-Asian racism emerging in Canada during the COVID-19 pandemic. Specifically, continuous struggles between decolonialization and colonialization in the postcolonial context are depicted through the decolonized multiculturalism policy that preserves colonization (Paquette et al., 2017). This policy provides a cultural environment where every cultural group has the benefit of equality, but this equality is in fact established and regulated by the dominant white English and French culture (MacDonald, 2014). These inequalities that run deep in Canadian society continue to influence intergroup interactions and become an important source of intergroup prejudice and discrimination (Jack-Davies, 2020; Roberto et al., 2020). The current devastation of the COVID-19 pandemic exposes these inequalities and scapegoats Asians for this pandemic (Bartholomew, 2020).

However, without clear sight of when or whether COVID-19 will be eradicated, the reemerging Anti-Asian racism may have more profound and lasting negative effects on the cultural structure and intergroup relations in Canada. This unprecedented situation requires us to adopt a more dynamic and reflective perspective to consistently review the Canadian social context, cultural policies, and intergroup interactions in order to better understand and manage the consequences of this pandemic. The significant role of social workers in this process has been integrated into this conceptual model. Specifically, social workers need to adopt a critical perspective to understand the role of the post-colonial context in contributing to the unequal cultural structure in Canadian society and how Canadian multiculturalism has normalized and legitimized this cultural structure. For instance, by educating themselves on Canadian multiculturalism policy, social workers can further reflect upon how the macro postcolonial context has shaped this policy and contributed to the unequal cultural hierarchy in Canada, a process that can become a concrete foundation for efforts to reduce anti-Asian racism in Canada. Particularly, by critically evaluating the Canadian context, social workers can better

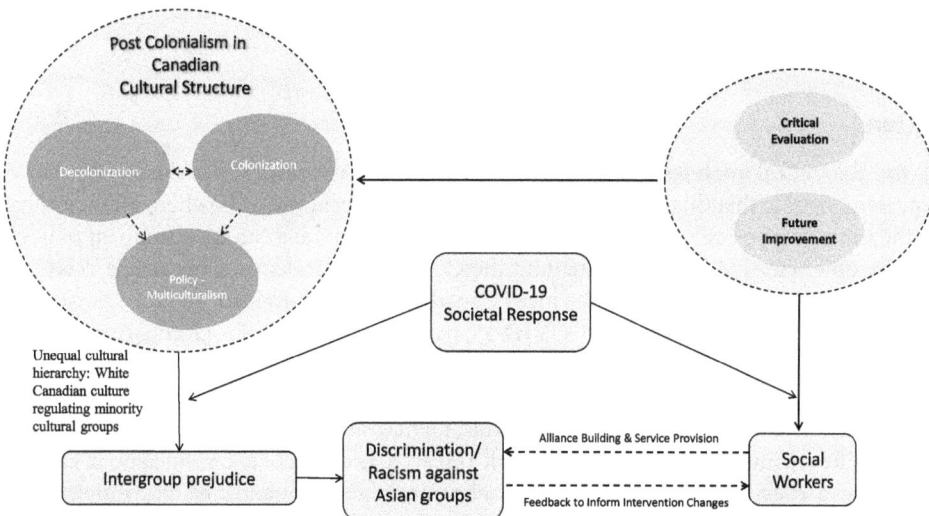

Figure 1. Conceptual model to understand anti-Asian racism in Canada amid COVID-19 and the roles of social workers.

understand and respond to the Asian community's circumstances regarding intergroup prejudice and discrimination during the COVID-19 pandemic. In addition, social workers can critically reflect on their own professional practice, evaluating the ways in which societal anti-Asian racist sentiments can unconsciously impact upon their personal social working behaviors. They can provide outreach to Asian communities during this difficult time and actively practice being anti-racist, seeking to better understand and bear witness to the lived experiences of racism amid the pandemic. These steps help social workers build a therapeutic alliance with Asian clients and pave the way for subsequent interventions and advocacy efforts. Through interactions with and feedback from these clients, targeted intervention programs can be improved and developed; furthermore, this information can be communicated to policymakers in order to continuously advance social policies and the social context. Specifically, social workers can collaborate with Asian individuals and communities throughout these procedures and integrate cultural sensitivity, diversity, and their lived experiences into intervention development and policy changes. Experiences and outcomes from these processes can further benefit a larger population of Black, Indigenous, and people of color during the pandemic and beyond.

This model can also be expanded upon in the future. For example, in this model, the Asian population is framed as a homogenous group. However, as there also exists diverse cultural groups (e.g., Chinese, Korean) within the larger Asian group, how subgroup dynamics and interactions are impacted by anti-Asian racism amid COVID-19 needs to further be developed. Within the Asian population, Chinese individuals are often the direct target of discrimination (Jack-Davies, 2020). Tessler et al. (2020) note that biased depictions provide the general public with a "straightforward narrative that focuses on China as the origin of COVID-19" (p. 637), leading other groups to place blame on Chinese for the accumulated fear and anger. Hence, how these depictions take place and what factors in the current context contribute to these depictions should be further investigated and included in future iterations of this conceptual model to help us better understand how anti-Asian racism is accelerated in the present day. Moreover, even though the imperative role of social workers in combating anti-Asian racism is exhibited in the conceptual model, more research is needed to confirm it. For instance, future directions mentioned above require social workers to thoroughly analyze contemporary Canadian context to detect more subtle development and effects of anti-Asian racism, a process that will be strengthened through further theoretical and empirical studies. Future research also needs to ascertain how social workers' critical analyses of the current Canadian social and cultural context can be transferred to practical knowledge and improve their interventions with Asian individuals experiencing discrimination. Future research also needs to focus on how social workers' experiences of dealing with anti-Asian racism can boost policy advocacy and facilitate cultural policy changes and social context reforms.

Conclusion

Overall, the theoretical analyses and conceptualization of anti-Asian racism during the COVID-19 pandemic demonstrate that this social phenomenon did not appear out of nowhere. Rather, it has solid roots in the Canadian postcolonial context, is maintained by Canadian multiculturalism policy, and is expressed by other cultural groups throughout the COVID-19 pandemic. This article delves into the macro postcolonial context in Canada, the meso Canadian multiculturalism, and the micro intergroup contacts, uncovering how the current COVID-19 pandemic has resurfaced anti-Asian racism and proposing the multiple and flexible roles of social workers in combating anti-Asian racism. Hence, this conceptualization provides a timely and integrated theoretical response to analyze this entrenched social phenomenon that results in unprecedented influences as the spread of COVID-19 continues. It can have significant implications for social work research and practice. Specifically, it can facilitate social workers to exercise their social justice leadership in the reduction of discrimination against Asian groups in Canadian society and leverage their values of social equality, diversity, and cohesion through research and practice. The experiences gained from this pandemic can provide important references to social workers in future crises and establish coping strategies to continuously promote social justice for underrepresented cultural groups, not only in Canadian society but in other multicultural societies around the world.

Disclosure statement

No potential conflict of interest was reported by the author(s).

Funding

The author(s) reported there is no funding associated with the work featured in this article.

ORCID

Kedi Zhao http://orcid.org/0000-0003-4847-673X

References

Abrams, D. (2010). *Processes of prejudices: Theory, evidence and intervention*. Equality and Human Rights Commission. https://www.equalityhumanrights.com/sites/default/files/research-report-56-processes-of-prejudice-theory-evidence-and-intervention.pdf

Ahlman, J. S. (2010). The Algerian question in Nkrumah's Ghana, 1958–1960: Debating "violence" and "nonviolence" in African decolonization. *Africa Today, 57*(2), 66–84. https://doi.org/10.2979/africatoday.57.2.66

Ahmed, S. (2000). *Strange encounters: Embodied others in post-coloniality*. Routledge. https://doi.org/10.4324/9780203349700

Arı, T., & Ak, M. (2019). Post-colonialism and international relations. In T. Ari & E. Toprak (Eds.), *Theories of international relations II* (pp. 67–86). Anadolu University Press.

Bartholomew, R. (2020, February 6). *The coronavirus and the search for scapegoats*. Psychology Today. https://www.psychologytoday.com/ca/blog/its-catching/202002/the-coronavirus-and-the-search-scapegoats

Bhatia, S., & Ram, A. (2001). Rethinking 'acculturation' in relation to diasporic cultures and postcolonial identities. *Human Development, 44*(1), 1–18. https://doi.org/10.1159/000057036

Bhuyan, R., Jeyapal, D., Ku, J., Sakamoto, I., & Chou, E. (2017). Branding 'Canadian experience' in immigration policy: Nation building in a neoliberal era. *Journal of International Migration and Integration, 18*(1), 47–62. https://doi.org/10.1007/s12134-015-0467-4

Burney, S. (2012). CHAPTER SEVEN: Conceptual frameworks in postcolonial theory: Applications for educational critique. *Counterpoints, 417*, 173–193. http://www.jstor.org/stable/42981704

Buzdugan, R., & Halli, S. S. (2009). Labor market experiences of Canadian immigrants with focus on foreign education and experience. *The International Migration Review, 43*(2), 366–386. https://doi.org/10.1111/j.1747-7379.2009.00768.x

Canadian Union of Public Employees. (2020, September 16). *COVID-19's impact on Indigenous, Black and racialized communities*. https://cupe.ca/covid-19s-impact-indigenous-black-and-racialized-communities

Cao, H., & Dehoorne, O. (2011). Changing territorial strategies: Chinese immigrants in Canada. In H. Cao & V. Poy (Eds.), *The China challenge* (pp. 222–240). University of Ottawa Press.

Christophers, B. (2007). Ships in the night: Journeys in cultural imperialism and postcolonialism. *International Journal of Cultural Studies, 10*(3), 283–302. https://doi.org/10.1177/1367877907080145

Chung, M. M. L. (2012). *The relationships between racialized immigrants and Indigenous peoples in Canada: A literature review*. [Master's thesis, Ryerson University]. Ryerson University Digital Repository.

Conway, S. (2018). From Britishness to multiculturalism: Official Canadian identity in the 1960s. *Études canadiennes/Canadian Studies, 84*(84), 9–30. https://doi.org/10.4000/eccs.1118

Croom, A. M. (2015). Slurs, stereotypes, and in-equality: A critical review of "How epithets and stereotypes are racially unequal." *Language Sciences, 52*, 139–154. https://doi.org/10.1016/j.langsci.2014.03.001

Das, M. (2020). Social construction of stigma and its implications – Observations from COVID-19. *SSRN Electronic Journal*. http://dx.doi.org/10.2139/ssrn.3599764

Day, R. J. F. (2000). *Multiculturalism and the history of Canadian diversity*. University of Toronto Press.

Dhanani, L. Y., & Franz, B. (2020). Unexpected public health consequences of the COVID-19 pandemic: A national survey examining anti-Asian attitudes in the USA. *International Journal of Public Health, 65*(6), 747–754. https://doi.org/10.1007/s00038-020-01440-0

Dmytriw, A. (2016). *Decolonizing immigration: Addressing missing indigenous perspectives in Canadian immigration policies.* [Master's thesis, Ryerson University]. Ryerson University Digital Repository.

Dovidio, J. F., & Gaertner, S. L. (1999). Reducing prejudice: Combating intergroup biases. *Current Directions in Psychological Science, 8*(4), 101–105. https://doi.org/10.1111/1467-8721.00024

Duckitt, J. (2003). Prejudice and intergroup hostility. In D. O. Sears, L. Huddy, & R. Jervis (Eds.), *Oxford handbook of political psychology* (pp. 559–600). Oxford University Press.

Enos, R. D., & Celaya, C. (2018). The effect of segregation on intergroup relations. *Journal of Experimental Political Science, 5*(1), 26–38. https://doi.org/10.1017/XPS.2017.28

Flanagan, R. (2020, July 8). *StatCan survey shows new evidence of increase in anti-Asian sentiment, attacks.* CTV News. https://www.ctvnews.ca/canada/statcan-survey-shows-new-evidence-of-increase-in-anti-asian-sentiment-attacks-1. 5016027

Gao, Z. (2020). Unsettled belongings: Chinese immigrants' mental health vulnerability as a symptom of international politics in the COVID-19 pandemic. *Journal of Humanistic Psychology, 61*(2), 198–218 doi:10.1177/ 0022167820980620.

Gee, G. C., Ro, M. J., & Rimoin, A. W. (2020). Seven reasons to care about racism and COVID-19 and seven things to do to stop it. *American Journal of Public Health, 110*(7), 954–955. https://doi.org/10.2105/AJPH.2020.305712

Gordon, J. (2020, December 15). *Black, minority women in Canada left behind in COVID-19 job recovery.* CTV News. https://www.ctvnews.ca/business/black-minority-women-in-canada-left-behind-in-covid-19-job-recovery-1. 5232390

Greensmith, C. D. (2015). *Diversity is (not) good enough: Unsettling White settler colonialism within Toronto's queer service sector.* [Doctoral dissertation, University of Toronto]. University of Toronto TSpace.

Haley, J. (2020, April 13). *COVID-19 and the threat to the North American economy.* Wilson Center. https://www. wilsoncenter.org/article/covid-19-and-threat-north-american-economy

Haque, E. (2012). *Multiculturalism within a bilingual framework: Language, race, and belonging in Canada.* University of Toronto Press.

Haslam, N., & Loughnan, S. (2014). Dehumanization and infrahumanization. *Annual Review of Psychology, 65*(1), 399–423. https://doi.org/10.1146/annurev-psych-010213-115045

Hiddleston, J. (2014). *Understanding postcolonialism.* Routledge.

Jack-Davies, A. (2020, August 3). *Coronavirus: The 'yellow peril' revisited.* The Conversation. https://theconversation. com/coronavirus-the-yellow-peril-revisited-134115

Jackson, H. (2021, March 22). *Canadian MPs vote to condemn Atlanta mass shooting, anti-Asian racism.* Global News. https://globalnews.ca/news/7712225/mps-condemn-anti-asian-hate/

Karamali, K. (2021, March 23). *Anti-Asian racism in Canada more 'frequent' as report tallies hundreds of attacks during pandemic.* Global News. https://globalnews.ca/news/7715260/anti-asian-racism-report-pandemic/.

Keil, R., & Ali, H. (2006). Multiculturalism, racism and infectious disease in the global city: The experience of the 2003 SARS outbreak in Toronto. *Topia: Canadian Journal of Cultural Studies, 16*, 23–49. https://doi.org/10.3138/topia.16. 23

Kibria, N. (2000). Race, ethnic options, and ethnic binds: Identity negotiations of second-generation Chinese and Korean Americans. *Sociological Perspectives, 43*(1), 77–95. https://doi.org/10.2307/1389783

Kumar, M. P. (2009). Aboriginal education in Canada: A postcolonial analysis. *AlterNative: An International Journal of Indigenous Peoples, 5*(1), 42–57. https://doi.org/10.1177/117718010900500104

Kustec, S. (2012). *The role of migrant labour supply in the Canadian labour market.* Citizenship and Immigration Canada. http://www.amssa.org/wp-content/uploads/2015/05/The-Role-of-migrant-labour-supply-in-the-Canadian-labour-market.pdf

Kymlicka, W. (1995). *Multicultural citizenship: A liberal theory of minority rights.* Oxford University Press. https://doi. org/10.1093/0198290918.001.0001

Kymlicka, W. (2004). Marketing Canadian pluralism in the international arena. *International Journal, 59*(4), 829–852. https://doi.org/10.2307/40203985

Kymlicka, W. (2013). Neoliberal multiculturalism? In P. Hall & M. Lamont (Eds.), *Social resilience in the neoliberal era* (pp. 99–126). Cambridge University Press. https://doi.org/10.1017/CBO9781139542425.007

Laurencin, C. T., & Walker, J. M. (2020). A pandemic on a pandemic: Racism and COVID-19 in blacks. *Cell Systems, 11* (1), 9–10. https://doi.org/10.1016/j.cels.2020.07.002

Law, H. K. T. (2018). *The changing faces of Chinese Canadians: Interpellation and performance in the deployment of the model minority discourse.* [Doctoral dissertation, York University]. York University York Space.

Lawrence, B., & Dua, E. (2005). Decolonizing antiracism. *Social Justice, 32*(4), 120–143. https://www.jstor.org/stable/ 29768340

Lee, J. M. (1999). *Asian minorities in Canada: Focusing on Chinese and Japanese people.* (Publication No. MQ45978) [Master's thesis, University of Toronto]. ProQuest Dissertations Publishing.

Lee, E. (2013). *A critique of Canadian multiculturalism as a state policy and its effects on Canadian subjects.* (Publication No. 1571139) [Master's thesis, University of Toronto]. ProQuest Dissertations Publishing.

Lee, S., & Waters, S. F. (2020). Asians and Asian Americans' experiences of racial discrimination during the COVID-19 pandemic: Impacts on health outcomes and the buffering role of social support. *Stigma and Health, 6*(1), 70–78 doi:10.1037/sah0000275.

Li, Y., & Nicholson, H. L., Jr. (2021). When "model minorities" become "yellow peril"—Othering and the racialization of Asian Americans in the COVID-19 pandemic. *Sociology Compass, 15*(2), e12849. https://doi.org/10.1111/soc4.12849

Little, W. (2016). *Introduction to sociology: 2nd Canadian edition.* BC Campus.

Loomba, A. (1998). *Colonialism/postcolonialism.* Routledge.

MacDonald, D. B. (2014). Aboriginal peoples and multicultural reform in Canada: Prospects for a new binational society. *Canadian Journal of Sociology, 39*(1), 65–86. https://doi.org/10.29173/cjs17224

Matthews, K. C. (2006). Perceiving discrimination: Psychological and sociopolitical barriers. *Journal of International Migration and Integration/Revue de L'integration Et de la Migration Internationale, 7*(3), 367–387. https://doi.org/10.1007/s12134-006-1018-9

McLaren, A. T., & Dyck, I. (2004). Mothering, human capital, and the "ideal immigrant." *Women's Studies International Forum, 27*(1), 41–53. https://doi.org/10.1016/j.wsif.2003.12.009

Molina, L. E., Tropp, L. R., & Goode, C. (2016). Reflections on prejudice and intergroup relations. *Current Opinion in Psychology, 11*, 120–124. https://doi.org/10.1016/j.copsyc.2016.08.001

Nagra, B., & Peng, I. (2013). Has multiculturalism really failed? A Canadian Muslim perspective. *Religions, 4*(4), 603–620. https://doi.org/10.3390/rel4040603

Nair, S. (2017, December 8). *Introducing postcolonialism in International Relations Theory.* E-International Relations. https://www.e-ir.info/2017/12/08/postcolonialism-in-international-relations-theory/

Oliver, J. E., & Wong, J. (2003). Intergroup prejudice in multiethnic settings. *American Journal of Political Science, 47*(4), 567–582. https://doi.org/10.2307/3186119

Paquette, J., Beauregard, D., & Gunter, C. (2017). Settler colonialism and cultural policy: The colonial foundations and refoundations of Canadian cultural policy. *International Journal of Cultural Policy, 23*(3), 269–284. https://doi.org/10.1080/10286632.2015.1043294

Pettigrew, T. F. (2008). Intergroup prejudice: Its causes and cures. *Actualidades en Psicología, 22*(109), 115–124. https://doi.org/10.15517/ap.v22i109.18

Pierce, D. (2009). Decolonization and the collapse of the British Empire. *Inquiries Journal, 1*(10). http://www.inquiriesjournal.com/articles/5/decolonization-and-the-collapse-of-the-british-empire

Pillay, T. (2015). Decentring the myth of Canadian multiculturalism. In A. A.a, L. Shultz, & T. Pillay (Eds.), *Decolonizing global citizenship education* (pp. 69–80). Sense Publishers. https://doi.org/10.1007/978-94-6300-277-6_6

Roberto, K. J., Johnson, A. F., & Rauhaus, B. M. (2020). Stigmatization and prejudice during the COVID-19 pandemic. *Administrative Theory & Praxis, 42*(3), 364–378. https://doi.org/10.1080/10841806.2020.1782128

Rowe, S., Baldry, E., & Earles, W. (2015). Decolonising social work research: Learning from critical Indigenous approaches. *Australian Social Work, 68*(3), 296–308. https://doi.org/10.1080/0312407X.2015.1024264

Sakamoto, I. (2007). A critical examination of immigrant acculturation: Toward an anti-oppressive social work model with immigrant adults in a pluralistic society. *British Journal of Social Work, 37*(3), 515–535. https://doi.org/10.1093/bjsw/bcm024

Schmid, P. C., & Amodio, D. M. (2017). Power effects on implicit prejudice and stereotyping: The role of intergroup face processing. *Social Neuroscience, 12*(2), 218–231. https://doi.org/10.1080/17470919.2016.1144647

Simpson, J. S., James, C. E., & Mack, J. (2011). Multiculturalism, colonialism, and racialization: Conceptual starting points. *Review of Education, Pedagogy, and Cultural Studies, 33*(4), 285–305. https://doi.org/10.1080/10714413.2011.597637

St. Denis, V. (2011). Silencing Aboriginal curricular content and perspectives through multiculturalism:"There are other children here." *Review of Education, Pedagogy, and Cultural Studies, 33*(4), 306–317 doi:10.1080/10714413.2011.597638.

Stephan, W. G., & Stephan, C. W. (2000). An integrated threat theory of prejudice. In S. Oskamp (Ed.), *Reducing prejudice and discrimination* (pp. 23–45). Erlbaum.

Tessler, H., Choi, M., & Kao, G. (2020). The anxiety of being Asian American: Hate crimes and negative biases during the COVID-19 pandemic. *American Journal of Criminal Justice, 45*(4), 636–646. https://doi.org/10.1007/s12103-020-09541-5

Thobani, S. (2007). *Exalted subjects: Studies in the making of race and nation in Canada.* University of Toronto Press.

Tsekouras, P. (2021, March 28). *Toronto's Asian community rallies against racism in wake of Atlanta spa shootings.* CTV News. https://toronto.ctvnews.ca/toronto-s-asian-community-rallies-against-racism-in-wake-of-atlanta-spa-shootings-1.5365699

Tuck, E., & Gaztambide-Fernandez, R. (2013). Curriculum, replacement, and settler futurity. *Journal of Curriculum Theorizing, 29*(1), 72–89. https://journal.jctonline.org/index.php/jct/article/view/411

Vernon, K. (2016). To the end of the hyphen-nation: Decolonizing multiculturalism. *ESC: English Studies in Canada, 42* (3–4), 81–98. https://doi.org/10.1353/esc.2016.0044

Viala-Gaudefroy, J., & Lindaman, D. (2020, April 21). *Donald Trump's 'Chinese virus': the politics of naming.* The Conversation. https://theconversation.com/donald-trumps-chinese-virus-the-politics-of-naming-136796

Wallis, C. (2020, June 12). *Why racism, not race, is a risk factor for dying of COVID-19.* Scientific American. https://www.scientificamerican.com/article/why-racism-not-race-is-a-risk-factor-for-dying-of-covid-191/.

Wang, Y. (2018). The cultural factors in postcolonial theories and applications. *Journal of Language Teaching and Research, 9*(3), 650–654. http://dx.doi.org/10.17507/jltr.0903.26

Wayland, S. V. (1997). Immigration, multiculturalism and national identity in Canada. *International Journal on Minority and Group Rights, 5*(1), 33–58. https://doi.org/10.1163/15718119720907408

World Health Organization. (nd.). *Coronavirus disease (COVID-19) pandemic.* https://www.euro.who.int/en/health-topics/health-emergencies/coronavirus-covid-19/novel-coronavirus-2019-ncov

Wu, C., Qian, Y., & Wilkes, R. (2020). Anti-Asian discrimination and the Asian-white mental health gap during COVID-19. *Ethnic and Racial Studies, 44*(5), 819–835 doi:10.1080/01419870.2020.1851739.

Young, R. J. C. (2016). *Postcolonialism: An historical introduction.* Wiley-Blackwell.

Young, I. (2020, January 29). *Those Asian people wearing face masks amid coronavirus fears? They aren't crazy, stupid or ridiculous, Vancouver. South China Morning Post.* https://www.scmp.com/news/china/article/3047964/those-asian-people-wearing-face-masks-amid-coronavirus-fears-they-arent

Zhou, S. (2020, June 22). *Almost one third of Chinese Canadians report being physically attacked during COVID-19.* ANTIHATE.CA. https://www.antihate.ca/_almost_one_third_of_chinese_canadians

Index

Note: Figures are indicated by *italics*. Tables are indicated by **bold**. Endnotes are indicated by the page number followed by 'n' and the endnote number e.g., 20n1 refers to endnote 1 on page 20.

For Product Safety Concerns and Information please contact our EU
representative GPSR@taylorandfrancis.com
Taylor & Francis Verlag GmbH, Kaufingerstraße 24, 80331 München, Germany

www.ingramcontent.com/pod-product-compliance
Lightning Source LLC
Chambersburg PA
CBHW081105220326
41598CB00038B/7235